ARTHUR GODMAN &
RONALD DENNEY

BARNES & NOBLE THESAURUS OF SCIENCE & TECHNOLOGY

BARNES & NOBLE BOOKS
A DIVISION OF HARPER & ROW, PUBLISHERS
New York, Cambridge, Philadelphia, San Francisco
London, Mexico City, São Paulo, Sydney

BLA Publishing Limited and the author would like to thank Rosie
Vane-Wright for her helpful advice and assistance in the production
of this book.

Library of Congress Cataloging-in-Publication Data

Godman, Arthur
    Barnes & Noble thesaurus of science & technology.

    1. Science--Dictionaries. 2. Technology--Dictionaries. I. Denney,
Ronald. II. Title. III. Title: Barnes and Noble thesaurus of science and
technology. IV. Title: Thesaurus of science & technology.
Q123.D33   1986   530'.21   85-45522
ISBN 0-06-463719-0 (pbk.)

This book was designed and produced by
BLA Publishing Limited, Swan Court,
London Road, East Grinstead, Sussex, England.

A member of the **Ling Kee Group**
LONDON·HONG KONG·TAIPEI·SINGAPORE·NEW YORK

Illustrations by Rosie Vane-Wright,

Phototypeset in Britain by BLA Publishing Limited and Composing
Operations Limited
Color origination by Planway Limited
Printed in Spain by Heraclio Fournier

Contents

BIOLOGICAL SCIENCES

How to use this book

This book combines the functions of a dictionary and a thesaurus: it will not only define a word for you, but it will also indicate other words related to the same topic, thus giving the reader easy access to one particular branch of the science. The emphasis of this work is on interconnections.

On pages 3 and 4 the contents pages list a number of broad groupings, sometimes with sub-groups, which may be used where reference to a particular theme is required. If, on the other hand, the reader wishes to refer to one particular word there is, at the back of the book, an alphabetical index in which approximately 2000 words are listed.

Looking up one particular word or phrase

Refer to the alphabetical index at the back of the book, then turn to the appropriate page. At the top of that page you will find the name of the general subject printed in bold type, and the specialised area in lighter type. For example, if you look up **constituent**, you will find it listed on p.45, at the top of which page is **INORGANIC CHEMISTRY**/COMPOUNDS. If you were unsure of the meaning of the phrase, you may now not only read its definition, but also place it in context. Immediately after the word or phrase you will see in brackets (parentheses) the abbreviation indicating which part of speech it is: (*n*) indicates a noun, (*v*) a verb and (*adj*) an adjective. Then follows a definition of the word, expressed as far as is possible in language which is in common use. Where a related word is listed nearby, a simple system using arrows has been devised.

(↑) means that the related term may be found above or on the opposite page.
(↓) means that the related term may be found below or on the opposite page.

A page reference in brackets is given for any word which is linked to the topic but is to be found elsewhere in the book. You will soon appreciate the advantages of this scheme of cross-referencing. Let us take an example. On p.163 the entry **erepsin** is:

erepsin (*n*) a mixture of enzymes (p.161) present in intestinal juice (↑) which complete the decomposition of protein (↓) to amino acids (↓) after the initial action of pepsin and trypsin (↑).

To gain a broader understanding, the reader will look at the entries **protein** and **amino acid** below on the same page, at the entry **intestinal juice** above on the same page, and will also refer to the entry **enzyme** on p.161.

Searching for associated words

As the reader will have observed, the particular organisation of this book greatly facilitates research into related words and ideas, and the extensive number of illustrations and diagrams assists in general comprehension.

Retrieving forgotten or unknown information

It would appear impossible to look up something one has forgotten or does not know, but this book makes it perfectly feasible. All that is required is a knowledge of the general area in which the word is likely to occur and the entries in that area will direct you to the appropriate word. If, for example, one wished to know more about an **aerofoil**, but had forgotten the term, it would be sufficient to know it was connected with **flight**; the reader looking up **flight** would be referred to **aerofoil**, which is defined and/or further explained by means of a diagram.

Studying or reviewing a subject

Two methods of using this book will be helpful to the reader who wishes to know more about a topic, or who wishes to review knowledge of a topic.

(*i*) For a broader understanding of seeds, for example, you would turn to the section dealing with this area and read through the different entries, following up the references which are given to guide you to related words.

(*ii*) If you have studied one particular branch of science and you wish to review your knowledge, looking through a section on **electronics**, by way of example, might refresh your memory or introduce an element which you had not previously realised was connected.

THE
THESAURUS

standard (*n*) a universally accepted unit of measurement, or basis of comparison. A standard measurement is very precisely defined, e.g. the metre is accepted as a standard of length and is measured as the wavelength of radiation from the chemical element krypton-86. The standard of mass is a piece of platinum-iridium alloy, i.e. an object of mass 1 kg, and is used as a basis of comparison for measuring the mass of other objects.

quantity (*n*) (1) any measurement of matter or energy is a quantity. Some examples of scientific quantities are mass, time, weight, density, length, temperature, concentration, heat, velocity, electric current and wavelength. (2) an unspecified, or indeterminable amount of a material or substance, e.g. the quantity of water in the beaker was unknown.

unit (*n*) a quantity accepted as a standard (↑) of measurement. The metre is a unit for measuring length; the second is a unit for measuring time: the pascal is a unit for measuring pressure.

magnitude[1] (*n*) size of a measurable quantity. Magnitude may be measured in different units (↑), e.g. 2 kg and 4.4 lb are different measurements of the same mass. They have the same magnitude.

variable (*adj*) describes a quantity that can change or be changed; it is not fixed, e.g. the temperature of a gas is variable, it can be increased or decreased. A variable resistor is one whose resistance to an electric current can be varied.

mean (*n*) a number derived by dividing the sum of two or more quantities by the number of quantities, e.g. the mean of the following temperatures, 9°C, 12°C, 8°C and 11°C is $(9 + 12 + 8 + 11)°C \div 4 = 40°C \div 4 = 10°C$. **mean** (*adj*).

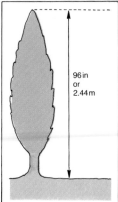

magnitude

96 in and 2.44 m are measurements of the same distance; they measure the same magnitude

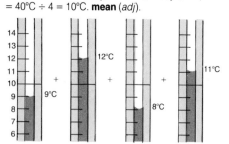

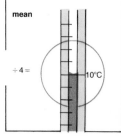

mean

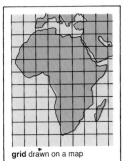

grid drawn on a map

grid (*n*) (↑) two sets of parallel lines, usually perpendicular to each other, drawn or marked on an area to provide a reference, e.g. grid lines on a map. (2) grid lines in the field of view of a microscope.

graph (*n*) a line plotted between two axes (lines) at right angles to each other. The line shows the mathematical relation between two variables (↑), e.g. a graph of mass against volume for a particular substance.

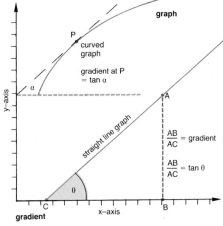

gradient (*n*) (1) the rate of change of a variable with distance, e.g. the change in height with distance along a slope; the change in potential with distance across an electric field. (2) the tangent of the angle made by a straight line graph with the x-axis. A straight line has a constant gradient. A curved graph has a varying gradient measured by the gradient of a tangent to the curve at any one point.

constant (*adj*) describes an unchanging quantity or measurement, e.g. an oven is kept at a constant temperature by a thermostat. **constant** (*n*).

approximate (*adj*) describes a measurement or number that is not absolutely accurate but is sufficiently close in magnitude (↑) to be used in calculations, e.g. the approximate value of π (which is good enough for most calculations) is 3.142.

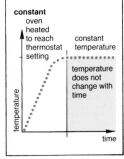

constant

mass (*n*) the quantity of matter in a body. It is not the
same as weight. The mass of a body always
remains the same, whereas its weight can vary
according to the force of gravity acting on it,
e.g. a man in space will have the same mass as he
has on Earth, but his weight will be less because the
force of gravity is less. All matter possesses mass; it
is measured in kilograms.

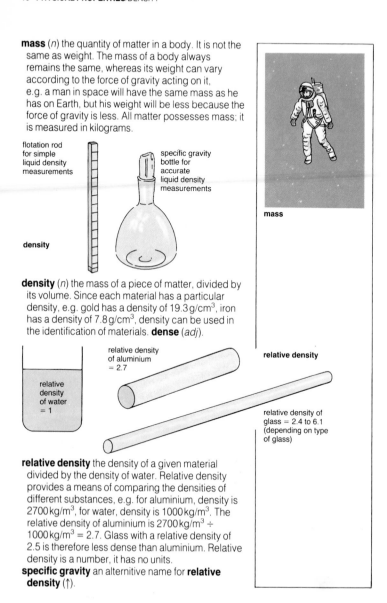

flotation rod
for simple
liquid density
measurements

specific gravity
bottle for
accurate
liquid density
measurements

density

mass

density (*n*) the mass of a piece of matter, divided by
its volume. Since each material has a particular
density, e.g. gold has a density of $19.3\,\text{g/cm}^3$, iron
has a density of $7.8\,\text{g/cm}^3$, density can be used in
the identification of materials. **dense** (*adj*).

relative density
of aluminium
= 2.7

relative density

relative
density
of water
= 1

relative density of
glass = 2.4 to 6.1
(depending on type
of glass)

relative density the density of a given material
divided by the density of water. Relative density
provides a means of comparing the densities of
different substances, e.g. for aluminium, density is
$2700\,\text{kg/m}^3$, for water, density is $1000\,\text{kg/m}^3$. The
relative density of aluminium is $2700\,\text{kg/m}^3 \div$
$1000\,\text{kg/m}^3 = 2.7$. Glass with a relative density of
2.5 is therefore less dense than aluminium. Relative
density is a number, it has no units.

specific gravity an alternitive name for **relative
density** (↑).

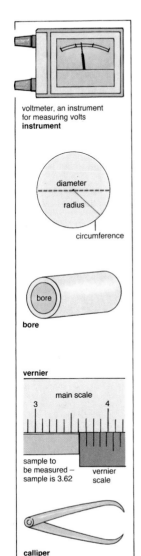

voltmeter, an instrument
for measuring volts
instrument

diameter

radius

circumference

bore

bore

vernier

main scale

3   4

sample to
be measured –
sample is 3.62   vernier
scale

calliper

device (*n*) an object constructed for a selected
special purpose to assist in work.

tool (*n*) a device which is used with the hands,
usually for construction or manufacturing, e.g. to
build offices, make cars, dig holes, repair shoes.

instrument (*n*) a device used for measuring or
recording information, or performing skilful work,
e.g. a voltmeter is an instrument used for measuring
an electric force, in volts.

gauge (*n*) a device used for measuring, or testing, in
engineering, meteorology, etc. The standard of
accuracy may be less than that of a measuring
instrument, e.g. a pressure gauge for measuring
pressure in boilers.

dimension (*n*) of a solid; its length, its height or its
breadth. Liquids and gases do not have dimensions
as they do not have an independent shape, but will
conform to the shape of their container. A line has
one dimension (length); a flat surface has two
dimensions (length and height); a solid has three
dimensions (length, height and breadth).

volume (*n*) the space taken up by a solid object,
liquid or gas, in three dimensions. A liquid or gas
has to be enclosed in a container to measure its
volume.

circumference (*n*) a curved line forming the outer
limit of a circle or a spherical object.

diameter (*n*) a straight line that passes from one side
of the circumference of a circle or sphere, through
the centre, to a point on the circumference at the
other side.

radius (*n*) the distance between the centre of a circle
or sphere and any point on the circumference.

bore (*n*) (1) the hollow part of a tube. (2) the internal
diameter of a tube.

scale (*n*) a numbered series of marks increasing
from a low value to a high value.

vernier (*n*) part of an accurate measuring instrument,
e.g. micrometer. A vernier is a short graduated
scale that slides along a larger scale to give a more
accurate measurement of, for example, length. In
the diagram the large scale shows a length of 3.6,
the vernier lines up at 2 so the accurate
measurement is 3.62.

calliper (*n*) an instrument for the accurate
measurement of small distances.

motion (*n*) movement; the act or process of going from one place to another, e.g. molecules are constantly in motion; the planets are in motion around the Sun. Motion occurs in either a straight line or a curved line.

speed (*n*) the rate at which motion occurs, measured as the distance travelled by an object divided by the time taken to travel that distance. The distance is not necessarily a straight line, and the direction of travel is not relevant to the calculation of speed.

velocity (*n*) speed in a straight line. Speed is a scalar quantity (p.22), velocity is a vector quantity (p.22). An object travelling along a curved path will have a constant speed if it travels equal distances in the same period of time; however, its velocity will be changing constantly as the direction in which it is travelling is changing constantly.

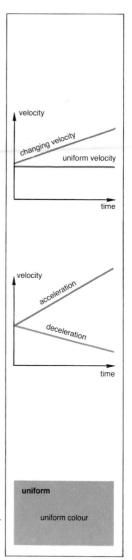

bike travelling at 12 metres per second

bike accelerates to 15 metres per second

acceleration

acceleration (*n*) the rate at which velocity is increased with time. If the velocity of a motor bike increases from 12 metres per second (m/s) to 15 m/s in 3 seconds, then the acceleration
= (increase in velocity) ÷ (time) = (12–15) m/s ÷ 3 s
= 3 m/s ÷ 3 s = 1 m/s^2 (metre per second per second). **accelerator** (*n*), **accelerate** (*v*), **accelerated, accelerating** (*adj*).

deceleration (*n*) the rate at which velocity is decreased with time. Deceleration is negative acceleration. **decelerate** (*v*).

rest (*n*) a state of not being in motion (↑), relative to an observer. A falling stone is in motion; once it has landed, it is at rest.

initial (*adj*) the first measurement or event to be considered, e.g. the initial speed of the motor car was 10 metres per second (m/s), though it accelerated to a speed of 50 m/s.

uniform (*adj*) describes a measurement which does not change with time, such as acceleration, velocity, speed. Describes any measurement or property which does not vary over a space, e.g. a solution with a uniform concentration; a crystal surface with a uniform colour.

velocity

changing velocity

uniform velocity

time

velocity

acceleration

deceleration

time

uniform

uniform colour

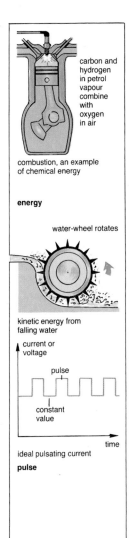

carbon and
hydrogen
in petrol
vapour
combine
with
oxygen
in air

combustion, an example
of chemical energy

energy

water-wheel rotates

kinetic energy from
falling water

ideal pulsating current

pulse

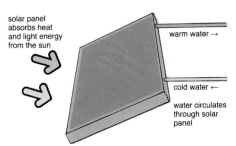

solar panel
absorbs heat
and light energy
from the sun

warm water →

cold water ←

water circulates
through solar
panel

energy (*n*) the ability to do work. Energy exists in
many forms: chemical energy, heat energy, light
energy, nuclear energy, mechanical energy either
as potential (stored) energy or kinetic (motion)
energy (p.14). Energy can be transformed from one
form to another. The symbol for energy is *E*.

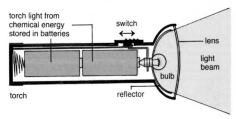

torch light from
chemical energy
stored in batteries

switch

lens

light
beam

bulb

torch

reflector

pulse (*n*) one variation in energy, becoming stronger
then returning to its original weaker strength,
e.g. a pulse of electric current rising to a maximum
then falling back to its original strength. **pulsate** (*v*).

joule (*n*) a unit of measurement of energy or work
(p.23). It is the work done when a force (p.20) of one
newton, N, displaces an object through a distance
of one metre in the direction of the force. The
symbol for joule is J.

power (*n*) the rate of doing work. It is measured as
the amount of work done divided by the time taken
to do it, e.g. if a machine does 30 J work in 6
seconds, the power of the machine is (30 J ÷ 6 s)
= 5 J/s. The symbol for power is *P*. The unit of power
is the watt (↓).

watt (*n*) a unit of measurement of power. It is the
power produced when one joule of work is done in
one second. The symbol for watt is W.

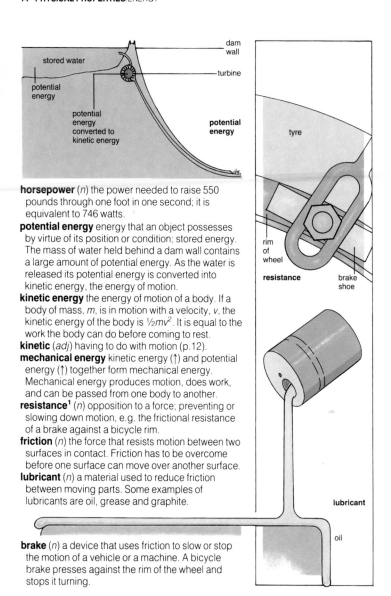

horsepower (*n*) the power needed to raise 550 pounds through one foot in one second; it is equivalent to 746 watts.

potential energy energy that an object possesses by virtue of its position or condition; stored energy. The mass of water held behind a dam wall contains a large amount of potential energy. As the water is released its potential energy is converted into kinetic energy, the energy of motion.

kinetic energy the energy of motion of a body. If a body of mass, *m*, is in motion with a velocity, *v*, the kinetic energy of the body is $\frac{1}{2}mv^2$. It is equal to the work the body can do before coming to rest.

kinetic (*adj*) having to do with motion (p.12).

mechanical energy kinetic energy (↑) and potential energy (↑) together form mechanical energy. Mechanical energy produces motion, does work, and can be passed from one body to another.

resistance[1] (*n*) opposition to a force; preventing or slowing down motion, e.g. the frictional resistance of a brake against a bicycle rim.

friction (*n*) the force that resists motion between two surfaces in contact. Friction has to be overcome before one surface can move over another surface.

lubricant (*n*) a material used to reduce friction between moving parts. Some examples of lubricants are oil, grease and graphite.

brake (*n*) a device that uses friction to slow or stop the motion of a vehicle or a machine. A bicycle brake presses against the rim of the wheel and stops it turning.

machine

spout

lever

when lever is pushed down with thumb, air pressure in oil can is increased forcing oil out of spout

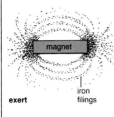

exert

magnet

iron filings

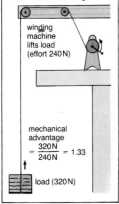

mechanical advantage

winding machine lifts load (effort 240N)

mechanical advantage
$= \dfrac{320\,N}{240\,N} = 1.33$

load (320N)

machine (n) a device that transfers energy; it uses a force applied at one point to overcome a force, usually at another point, e.g. a lever; a system of pulleys. In general, a device that transforms energy; it takes in some definite form of energy, modifies it and delivers the energy in a suitable form for a desired purpose, e.g. a motor car transforms chemical energy into kinetic energy; a turbine (p.19) and a generator transform kinetic energy from water, or some other fluid, into electrical energy. Compare **engine** (p.17).

effort (n) the force applied to any machine to do work.

exert (v) to apply a force, or initiate an action, e.g. a magnet exerts an attractive force on iron filings.

load (n) (1) the weight raised or moved by a lever, or other machine. (2) the weight carried by beams and pillars in buildings.

input (n) the energy, or work, that is put into a machine (↑).

output (n) the power or work that is obtained from a machine (↑).

mechanical advantage the number of times the load raised or moved by a machine is greater than the applied force (effort), e.g. a load of 320N is raised by an effort of 240N; the mechanical advantage of the machine is (320N ÷ 240N) = 1.33. The higher the mechanical advantage of a machine, the greater its efficiency (↓).

velocity ratio of a machine the ratio of the distance moved by the effort to the corresponding distance moved by the load.

ratio (n) a comparison of two numbers or of two values with the same units, e.g. the ratio of 20m to 80m is 1:4; the ratio of 8.1kg to 0.9kg is 9:1.

efficiency (n) the ratio of the useful work done by a machine to the total work put in. It is usually expressed as a percentage. If 320J of work is done by the effort and 240J of work is done on the load, then the efficiency = (240J ÷ 320J) × 100% = 75%. No machine has 100% efficiency (i.e. no machine is perfect) since some work is always lost in overcoming friction between moving parts of a machine. Lubricants (↑) are used to reduce friction between moving parts of a machine and so improve efficiency.

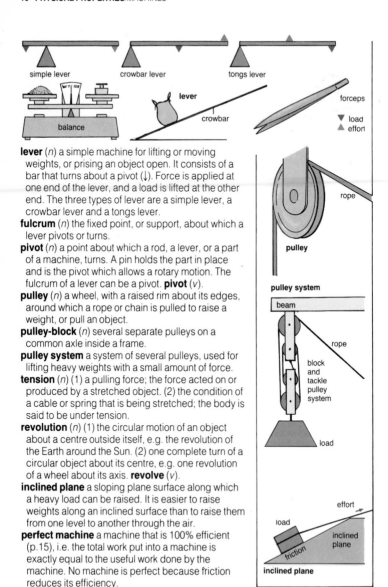

simple lever   crowbar lever   tongs lever

balance

lever

crowbar

forceps

▼ load
▲ effort

rope

pulley

pulley system

beam

rope

block
and
tackle
pulley
system

load

effort

load

inclined
plane

friction

inclined plane

lever (*n*) a simple machine for lifting or moving weights, or prising an object open. It consists of a bar that turns about a pivot (↓). Force is applied at one end of the lever, and a load is lifted at the other end. The three types of lever are a simple lever, a crowbar lever and a tongs lever.

fulcrum (*n*) the fixed point, or support, about which a lever pivots or turns.

pivot (*n*) a point about which a rod, a lever, or a part of a machine, turns. A pin holds the part in place and is the pivot which allows a rotary motion. The fulcrum of a lever can be a pivot. **pivot** (*v*).

pulley (*n*) a wheel, with a raised rim about its edges, around which a rope or chain is pulled to raise a weight, or pull an object.

pulley-block (*n*) several separate pulleys on a common axle inside a frame.

pulley system a system of several pulleys, used for lifting heavy weights with a small amount of force.

tension (*n*) (1) a pulling force; the force acted on or produced by a stretched object. (2) the condition of a cable or spring that is being stretched; the body is said to be under tension.

revolution (*n*) (1) the circular motion of an object about a centre outside itself, e.g. the revolution of the Earth around the Sun. (2) one complete turn of a circular object about its centre, e.g. one revolution of a wheel about its axis. **revolve** (*v*).

inclined plane a sloping plane surface along which a heavy load can be raised. It is easier to raise weights along an inclined surface than to raise them from one level to another through the air.

perfect machine a machine that is 100% efficient (p.15), i.e. the total work put into a machine is exactly equal to the useful work done by the machine. No machine is perfect because friction reduces its efficiency.

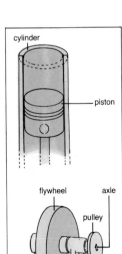

cylinder

piston

flywheel   axle

pulley

flywheel

engine (*n*) a machine that transforms heat and other forms of energy into mechanical energy, e.g. the engine of a motor car uses heat energy, generated by the ignition of a mixture of petrol and air, and transforms it into mechanical energy.

cylinder (*n*) part of an engine in which a piston moves up and down. Vapour is sucked in through one hole in the cylinder, and leaves through another hole after it has been expanded.

piston (*n*) a circular disk, or cylindrical piece of metal that moves up and down in a tightly fitting cylinder. The piston moves up to compress the vapour in the cylinder, and is forced down when the vapour expands. The movement of the piston produces mechanical energy.

flywheel (*n*) a heavy wheel rotating on a shaft (p.33); it stores kinetic energy due to its speed of rotation. An engine using a piston to supply energy, only supplies energy for part of its working cycle. It supplies energy to the flywheel, the energy is stored in the flywheel and then is returned to the piston to keep the piston moving during non-working strokes.

steam engine a type of engine that uses steam under pressure to drive a piston and produce mechanical energy. Heat energy boils water to produce steam under pressure. The steam expands in a cylinder, pushing the piston down. As the expanded steam escapes through the exhaust (p.18), the piston rises. The piston is attached to a crank which changes the up and down motion to a circular one.

safety valve an outlet device which operates a valve so that it opens when the steam pressure in a boiler reaches a particular value, the maximum pressure allowed in the boiler. When the pressure falls, the valve closes and steam no longer escapes.

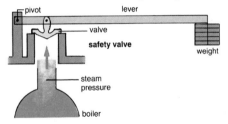

pivot   lever

valve

safety valve

weight

steam pressure

boiler

reciprocating engine an engine in which the movement of a piston up and down in a cylinder produces the rotation of a wheel by means of a crankshaft (p.33). The action of the crankshaft returns the piston to its up position.

internal combustion engine an engine in which a mixture of liquid fuel and air is ignited inside a cylinder to drive a piston. In a four-stroke engine the following occurs: at stroke (1) a mixture of fuel and air is sucked into the cylinder; at stroke (2) the mixture is compressed; at stroke (3) the mixture is ignited and it explodes, forcing the piston down; at stroke (4) the exhaust gases are forced out of the cylinder.

internal combustion engine

1. petrol vapour sucked into cylinder
2. vapour compressed
3. spark ignites mixture of petrol and air
4. exhaust gases pushed out of cylinder

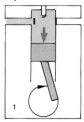 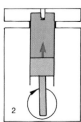 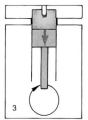 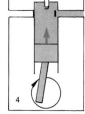

ignition (*n*) the explosion of the mixed gases in an internal combustion engine (↑), or the parts which bring about the explosion. An alternator (p.80) produces a high voltage which forms a spark (p.82) between the points of a **sparking plug**. The spark explodes the gaseous mixture in the cylinder at the correct time.

exhaust (*n*) (1) a flow of gases which are no longer required by an engine (p.17), e.g. the products of combustion in an internal combustion engine (↑) form the exhaust when they leave the cylinder. (2) the pipe along which the waste exhaust gases flow. **exhaust** (*v*).

transmission (*n*) the physical means, or the process, by which power from an engine is taken to the place where it is used, e.g. the transmission of a motor car takes power from the engine to the road wheels. Shafts and gear wheels are a form of transmission.

Otto cycle the four-stroke cycle of an internal combustion engine (↑). The strokes are: induction; compression; ignition; exhaust.

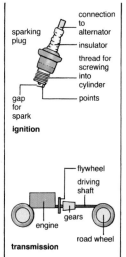

sparking plug
connection to alternator
insulator
thread for screwing into cylinder
gap for spark
points

ignition

flywheel
driving shaft
gears
engine
road wheel

transmission

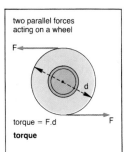

two parallel forces
acting on a wheel

F

d

torque = F.d   F

torque

torque (*n*) a force, or a moment of a force, which twists or turns a rod or shaft, or turns a wheel. A torque tends to produce rotation; if rotation is produced, work is done. The quantity of work depends on the torque and the total angle through which the rod, shaft, or wheel has rotated. Work is measured in joules, the angle in radians and the torque in newton metres:

Work done = torque × angle rotated

traction (*n*) the action of pulling, usually applied to an engine pulling a load on a surface which offers frictional resistance. The engine overcomes **tractive** resistance which consists of mechanical (frictional) resistance and air resistance, the latter being more important at high speeds. The tractional pull of an engine is measured in newtons.

turbine (*n*) a type of engine in which a current of air, steam, water or other fluid is directed against propeller blades attached to a shaft. The force of the current turns the blades, which turn the shaft.

jet engine liquid fuel, such as kerosene (p.120) is mixed with air and burnt in a combustion chamber. The hot gaseous products expand and produce a jet of exhaust gases; the jet of gases pushes the engine forward by jet propulsion (↓). The jet also turns turbines which drive a compressor. The compressor pulls in air, compresses it, and passes it to the combustion chamber.

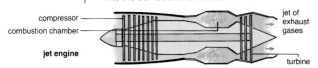

compressor
combustion chamber

jet of
exhaust
gases

jet engine

turbine

thrust (*n*) a force which pushes. It is the propulsive force produced by a jet engine (↑) and it is the force with which one part of a solid structure pushes against another part (in this case there is no motion). **thrust** (*v*).

jet propulsion the forward propulsion of an aircraft or boat caused when very hot gases, under great pressure, are released through an opening at the back of the jet engine of the vehicle. The thrust of the gases backwards produces an equal and opposite reaction that propels the vehicle forwards.

reaction propulsion The same as **jet propulsion**.

momentum (*n*) the product of the mass (p.10) and the velocity of a body, e.g. an object with a mass of 10kg moving at a velocity of 6m/s, has a momentum of 60kgm/s. All solids and fluids in motion possess momentum.

force (*n*) an external agency capable of starting or stopping the motion of an object, changing its direction when it is already moving, or changing its shape. A force is not visible, only its effect can be seen and measured. Some common forces are gravity, flowing water, wind and magnetism. A force can be measured as the change in momentum it produces in 1 second; or the amount it extends a spring; or the acceleration it gives to a mass. Force = (mass) × (acceleration), expressed in symbols as $F = ma$.

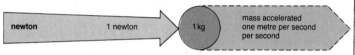

newton | 1 newton | 1 kg | mass accelerated one metre per second per second

newton (*n*) the unit of force (↑). One newton is the force required to give a mass of one kilogram an acceleration of one metre per second per second. The symbol is N.

action (*n*) the effect of a force, e.g. the action of a hammer hitting a nail will be the effect of the force of the hammer, i.e. the nail penetrating the wood. Every action has an equal and opposite reaction (↓). **act** (*v*).

action

reaction (*v*) the equal and opposite force produced by an action (↑), e.g. when a ball hits a wall it has an action on the wall. The wall does not change its shape, but it pushes back with a reaction that is equal and opposite to the action of the ball.

inertia (*n*) the tendency of an object to stay in the same state, whether at rest or continuing to move at a uniform rate in a straight line. A force is needed to overcome inertia.

reaction

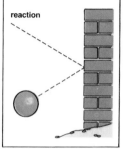

centrifugal (*adj*) trying to move away from a centre of rotation, e.g. centrifugal force is the reaction of a body travelling in a circular path, against the force keeping it in the path.

centripetal (*adj*) tending to move towards a centre of rotation, e.g. centripetal force is the force which keeps a body travelling in a circular path. If the tension in a string keeps a body in motion at a constant linear velocity in a circle, then the tension is equal to the centripetal force. If a body of mass m travels in a circle of radius r at a constant linear velocity v, then the centripetal force is mv^2/r.

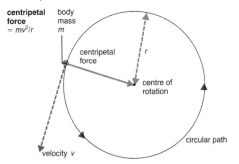

centripetal force = mv^2/r body mass m centripetal force r centre of rotation circular path velocity v

attraction (*n*) a force that tends to draw two or more objects towards one another. Gravity and magnetism are forces of attraction.

gravity (*n*) the force of attraction exerted by the Earth, Moon or other bodies in space on all solids, liquids and gases, e.g. a ball thrown up into the air will fall to the ground; it is being acted on by gravity. Gravity gives people and objects weight. Gravity is dependent on two factors; the mass of the objects concerned, and their distance apart. The greater the mass, the greater the gravitational attraction; the greater the distance apart, the smaller the gravitational attraction.

weight (*n*) the gravitational force which attracts any mass towards the Earth, the Moon or other bodies in space. The weight of an object is measured in newtons N, and varies with the strength of the force of gravity. The greater the force of gravity acting on an object, the greater its weight. **weigh** (*v*) to measure the weight of an object. *See* **mass**.

attraction

magnetic flux lines showing attraction around a bar magnet

weight

$1\,kg \equiv 9.83\,N$ measured at north pole

equator

Earth

$1\,kg \equiv 9.78\,N$ measured at equator

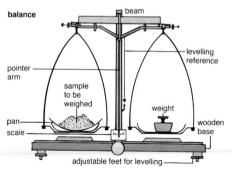

balance

beam

levelling reference

pointer arm

sample to be weighed

weight

pan

scale

wooden base

adjustable feet for levelling

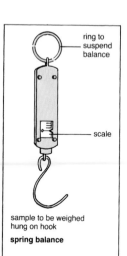

ring to suspend balance

scale

sample to be weighed hung on hook

spring balance

balance (*n*) (1) an apparatus for measuring mass (p.10). Masses placed in the scale pans are subject to gravity. When the force acting on the masses is equal, as shown when the arm of the balance is level, the masses themselves must be equal. (2) a condition of partial or complete equilibrium.

spring balance a device for measuring weight. The weight of an object suspended on the spring extends the spring, and a pointer indicates the object's weight on a scale.

vector[1] (*n*) a quantity that has a direction and a magnitude, and a full description must include both, e.g. velocity is a vector quantity; it is speed (magnitude) in a straight line (direction). A vector quantity can be represented by a straight line; its direction shows the direction of the vector, its length the magnitude; and its origin the point of action.

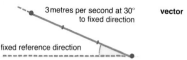

3 metres per second at 30° to fixed direction

vector

fixed reference direction

scalar (*n*) a quantity that has magnitude but no direction, e.g. speed is fully defined as the distance travelled by an object divided by the time taken to travel that distance. It can be understood without any reference to direction.

resultant (*n*) a single force that could replace two or more forces acting on an object, without altering the effect.

component[1] (*n*) one of two or more forces that produce the same effect on a body as a single force.

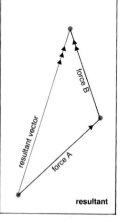

force B

resultant vector

force A

resultant

equilibrium (*n. pl. equilibria*) the state of a body when it is at rest under the action of opposing forces. Forces which balance each other are in equilibrium.

stable equilibrium an object is in stable equilibrium if it returns to its original position after being tipped slightly, e.g. a cone on a table.

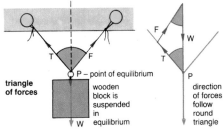

triangle
of forces

P – point of equilibrium

wooden
block is
suspended
in
equilibrium

direction
of forces
follow
round
triangle

unstable equilibrium an object is in unstable equilibrium if it moves further away from its original position after being tipped slightly, e.g. a cone balancing on its point.

neutral equilibrium an object is in neutral equilibrium if it stays in its new position after being moved slightly, e.g. a ball on a level floor.

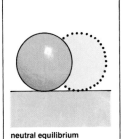

neutral equilibrium

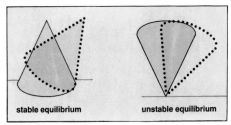

stable equilibrium   unstable equilibrium

triangle of forces if three forces can be represented in magnitude and direction by the sides of a triangle, taken in order, then the three forces are in equilibrium.

work (*n*) the result when a force moves an object. Work is done whenever an object is moved, or its motion stopped by a force. It is measured as the size of the force multiplied by the distance moved in the direction in which the force acts. The symbol for work is w. The unit of work is the joule.

molecule (*n*) the smallest particle of an element or compound capable of existing independently and still retaining the chemical properties of that element or compound. It consists of atoms (p.45). Molecules may be held together by means of attractive forces.

cohesion (*n*) the force of attraction between molecules of one substance that makes them hold together, e.g. the molecules in mercury are held together by cohesion. **cohesive** (*adj*), **cohere** (*v*).

adhesion (*n*) the force of attraction between molecules of different substances that makes them hold together, e.g. forces of adhesion make water stay attached to the surface of glass. **adhesive** (*adj*), **adhere** (*v*).

surface tension the property of a liquid whereby its surface behaves as if it is covered by a thin, elastic skin, e.g. a drop of mercury on a flat surface is spherical, or near spherical, rather than flat because of surface tension. It occurs because molecules at or near the surface of a liquid are acted upon by cohesive forces in the liquid, but not above the surface. The surface is under a state of tension which tends to draw it together so that it has the smallest area. In the case of small quantities of liquid, this shape is a sphere.

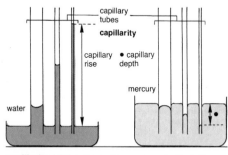

capillarity (*n*) the rise or depression of a liquid in a capillary tube due to surface tension. A capillary tube is a tube with a very narrow bore. The smaller the diameter of the capillary tube the greater the rise or depression of the liquid.

meniscus (*n*) the curved surface of a liquid enclosed in a narrow tube; it may be concave or convex.

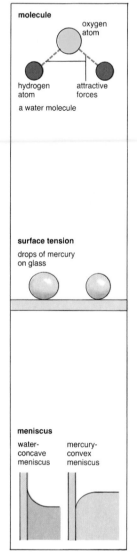

molecule

oxygen atom

hydrogen atom

attractive forces

a water molecule

surface tension

drops of mercury on glass

meniscus

water-concave meniscus

mercury-convex meniscus

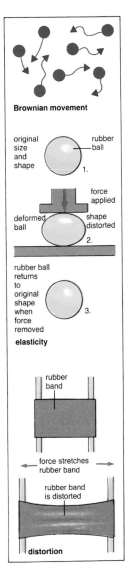

Brownian movement

original size and shape

rubber ball

1.

force applied

deformed ball

shape distorted

2.

rubber ball returns to original shape when force removed

3.

elasticity

rubber band

force stretches rubber band

rubber band is distorted

distortion

Brownian motion or **movement** the random, continuous motion of microscopic particles in a liquid or gas; it is caused by the impact of the constantly moving molecules of the gas or liquid on the particles.

diffusion (*n*) the process by which molecules of a substance spread through a solid, a liquid or a gas. Diffusion stops when the molecules are spread evenly throughout. A soluble crystal placed in water slowly diffuses throughout the whole of the water. Diffusion is brought about by the continuous impact of moving molecules of solvent.

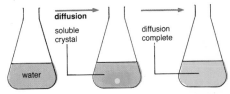

diffusion

soluble crystal

diffusion complete

water

property (*n*) any one of the physical or chemical characteristics of a material that may assist in its identification. Examples of the physical properties of substances are melting and boiling points, colour and shape. Chemical properties are exhibited in the way one substance reacts with other substances.

elasticity (*n*) the property of an object, or material, such that it returns to its original size, or shape, after a force applied to it has been removed; all substances possess some degree of elasticity. **elastic** (*adj*) rubber is highly elastic.

extension (*n*) the increase in length of an elastic solid when it is stretched by a force. **extend** (*v*).

distortion[1] (*n*) the process in which the shape of an object is temporarily changed, e.g. the change in shape when a piece of rubber is stretched or squeezed. Elastic (↑) materials are distorted as the change in shape is only temporary. **distort** (*v*), **distorted** (*adj*).

deformation (*n*) (1) a change in the size or shape of a solid. An elastic solid within the elastic limit (p.26) returns to its original size or shape when the deforming force no longer acts. The deformation of a plastic solid remains when the force is removed. (2) the amount by which an object is changed from its original state. **deform** (*v*), deformed (*adj*).

stress (*n*) a force per unit area acting on a solid. A deformation (p.25) or distortion of the solid is produced by the stress.

strain (*n*) a change in the size or shape of a solid when a stress is applied. Strain is measured by: (change in dimension) ÷ (dimension before stress); the dimension can be length, area or volume. For a particular material, (stress) ÷ (strain) is a constant within the elastic limit (↓).

Hooke's law the extension of an elastic solid is proportional to the force stretching it, providing the force is below the elastic limit (↓).

limit (*n*) a point, moment in time, line or magnitude which cannot be exceeded, or beyond which it is not possible to go, e.g. the limit of absolute zero of temperature.

elastic limit the limit of force below which a solid will return to its original length after having been stretched by a force, and above which the solid will not return to its original state. Beyond the elastic limit, the solid will no longer be elastic.

restoring force the force within a body which returns or tends to return it to its former position or original shape, e.g. when an elastic band is extended by a force, a restoring force returns the band to its original length.

rigid (*adj*) describes a material which cannot easily be bent or deformed when acted on by a force.

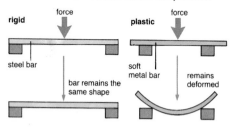

plastic¹ (*adj*) describes a material that alters its shape or size when acted on by a force, and does not return to its original shape when the force is removed, e.g. clay.

brittle (*adj*) describes a solid which breaks into pieces when a force is applied, e.g. glass is brittle. **brittleness** (*n*).

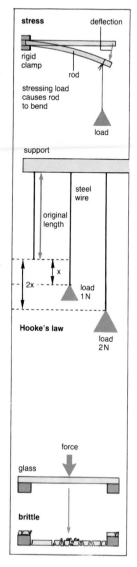

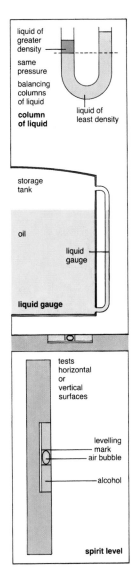

liquid of greater density

same pressure

balancing columns of liquid

column of liquid

liquid of least density

storage tank

oil

liquid gauge

liquid gauge

tests horizontal or vertical surfaces

levelling mark

air bubble

alcohol

spirit level

pressure (*n*) a pushing, squeezing force. It is measured as the strength of a force acting on a surface, divided by the area over which that force acts, i.e. as units of force per unit area. Force is measured in newtons, N, and area in square metres, m², so pressure is measured in newtons per square metre, N/m². The symbol for pressure is *p*.

pascal (*n*) the unit of pressure, equal to one newton per square metre. The symbol is Pa.

column of liquid liquid contained in a tube. The height of the liquid and its density determine the pressure on the bottom of the tube. The pressure is larger the greater the height of the liquid, or the greater the density of the liquid. Columns of different liquids can be balanced and their relative densities calculated.

low pressure   **column of liquid**

pressure

high pressure

liquid gauge a device, external to a vessel, used to measure the depth of liquid in the vessel, e.g. oil in a storage tank.

spirit level a device used to test whether a surface is exactly horizontal or exactly vertical. It contains a bubble of air, or oil, in a liquid (usually alcohol); the bubble will lie between two marks if the surface is exactly horizontal or vertical.

Pascal's law pressure applied to any part of an enclosed liquid is transmitted, without diminishing, in all directions, to every other part of the liquid.

hydraulic (*adj*) operated by the pressure or movement of a liquid, e.g. a hydraulic jack. A jack has one large and one small cylinder joined by a pipe. The pressure in a liquid is the same in both cylinders. Pressure equals force ÷ area, expressed in symbols, $p = FA$, so this produces a high pressure in this cylinder. In the large cylinder, force = pressure × area, so a large force is produced by the pressure in this cylinder, as the pressure in both cylinders is the same. A small force (effect) has produced a very large force, capable of lifting heavy weights.

streamline (*adj*) refers to the flow of a liquid, or gas, in which there is no sudden change of direction.

streamlined (*adj*) describes the shape of a surface such that any fluid passing over it will have a streamline flow.

turbulent (*adj*) describes the flow of a fluid in which the flow changes continually and irregularly in magnitude and direction. It is the opposite of streamline (↑).

suspend (*v*) (1) to hang from a support so as to allow free movement, e.g. to suspend a load by a rope from a hook or crane, (2) to hold or to keep in a position, e.g. particles suspended in a liquid.

immerse (*v*) to place a solid object partially, or wholly, under the surface of a liquid.

submerge (*v*) to immerse (↑) a solid object completely under water, usually at some depth, e.g. a submarine submerges below the surface.

displacement (*n*) (1) the weight or volume of fluid pushed out of place when an object is immersed (↑). (2) the difference between the original position of an object and any subsequent position. **displaced** (*adj*), **displace** (*v*).

measuring cylinder a tall glass vessel used to measure the volume of a liquid, commonly graduated in cubic centimetres.

Archimedes' principle when an object is wholly, or partially, immersed (↑) in a liquid, there is an apparent loss in weight, this loss is caused by an upthrust (↓) by the liquid on the object and is equal to the weight of liquid displaced. Archimedes' principle can be applied to gases as well as liquids.

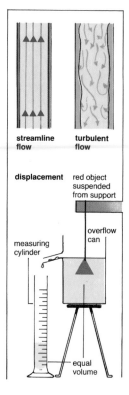

streamline flow   turbulent flow

displacement   red object suspended from support

overflow can

measuring cylinder

equal volume

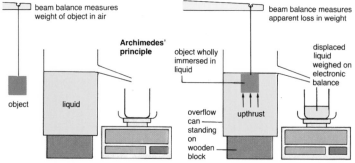

beam balance measures weight of object in air

beam balance measures apparent loss in weight

Archimedes' principle

object wholly immersed in liquid

displaced liquid weighed on electronic balance

object   liquid

overflow can standing on wooden block

upthrust

reading here

scale

test
liquid

hydrometer

weight

balloon

**Venturi tube
(Bernouilli's principle)**

upthrust (*n*) a force acting in a vertical upward
direction.

flotation (*n*) the act of floating. An object of weight,
W, floating in a fluid displaces an equal weight, W,
of that fluid. This law of flotation is a special case of
Archimedes' principle (↑). **float** (*v*).

buoyancy (*n*) (1) the ability of an object to float in a
fluid. The magnitude of buoyancy determines
whether an object sinks or floats. (2) the upthrust of
a fluid on an immersed object. **buoyant** (*adj*).

hydrometer (*n*) a floating glass or metal instrument
(p.11) used for measuring the density (p.10) of a
liquid. It has a weight at the base to make it float
upright and a graduated stem. The lower the
density of the liquid, the lower the hydrometer sinks.

balloon (*n*) a flexible bag filled with hot air or a gas of
low density (p.10). The balloon displaces air and
will rise if the upthrust of the displaced air is greater
than the weight of the balloon.

Boyle's law states that the volume of a confined gas
at a constant temperature varies inversely with the
pressure, i.e. if the pressure on the gas is
increased, there will be a proportionate decrease in
volume; (pressure of gas) × (volume of gas)
= constant. Boyle's Law is represented in symbols
as $pV = k$ (a constant).

Bernouilli's principle when a liquid flows through a
horizontal tube, the amount of energy per kilogram
of the liquid does not change. The energy is the
product of the pressure and velocity of the liquid. If
velocity increases (as it does if the tube narrows),
then the pressure of the liquid falls; if velocity
decreases, then the pressure increases.

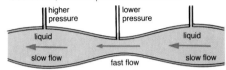

higher
pressure

lower
pressure

liquid

liquid

slow flow

fast flow

slow flow

Venturi tube a tube which can be used to illustrate
Bernouilli's principle (↑). The tube narrows in its
middle section. Manometers (p.32) attached to the
wide and narrow sections of the tube indicate the
pressure of the fluid in these sections. The pressure
is low in the narrow section of the tube, and higher
in the wider sections.

hull (*n*) the body of a boat.

keel (*n*) a part of a boat, along the bottom of the hull
(↑), extending from the front to the back. It is very
heavy and its purpose is to keep a boat in an
upright position when wind or waves push the boat
over to one side.

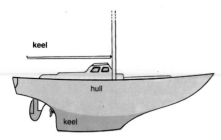

hydrofoil (*n*) (1) a flat, or curved structure attached
to the hull (↑) of a boat. It acts on the water and
raises the hull above the water when the boat
travels fast. (2) a boat with hydrofoils; it can travel
much faster than an ordinary boat.

hydroplane (*n*) (1) a motor boat with a flat hull (↑) and
no keel (↑); the shape allows the boat to rest on the
surface of water, and to travel at a fast speed. (2) a
fin-like structure which stabilizes a submarine when
submerged.

hovercraft (*n*) a type of boat which is supported on a
cushion of air. The air is blown down by a fan and
contained by a leather skirt attached to the hull (↑).
Propulsion is provided by air propellors and the
hovercraft can move over water and land.

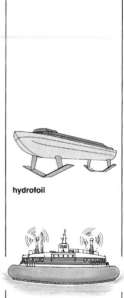

hydrofoil

hovercraft

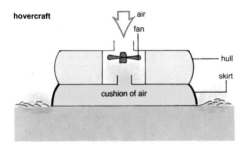

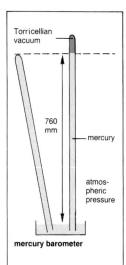

mercury barometer

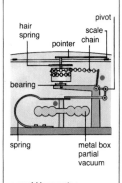

aneroid barometer

● movements of box
amplified by
lever system

atmospheric pressure the pressure exerted by the
atmosphere, not only downwards on the Earth's
surface, but in all directions. Normal pressure is
taken as 101 325 pascals. This will support a
column of mercury 760mm high. Atmospheric
pressure varies around this value from day to day.

bar (*n*) a unit of pressure for measuring atmospheric
pressure. One bar = 10^5 newtons per square metre
= 750.076 mm mercury pressure.

barometer (*n*) an instrument that measures
atmospheric pressure (↑). There are several
different types of barometer.

mercury barometer a barometer that consists of
mercury in a tube sealed at the top, with the bottom
immersed in a vessel containing mercury. When
atmospheric pressure is high, the pressure on the
surface of the mercury increases and the level of
mercury in the tube rises. When atmospheric
pressure is low, the pressure on the surface of the
mercury decreases and the level of mercury in the
tube falls.

absolute vacuum a space from which all of the gas
or air has been removed. In practice it is not
possible to create an absolute vacuum.

vacuum (*n*) a space from which as much gas or air
as possible has been taken is called a **high
vacuum**; a space from which some gas or air has
been taken is called a **low** or **partial vacuum**.

Torricellian vacuum the name given to the vacuum
above the column of mercury in a barometer.

aneroid barometer a barometer that indicates
atmospheric pressure, the pressure exerted by the
atmosphere, by the expansion and contraction of a
metal box. Air has been removed from the box,
creating a partial vacuum. When atmospheric
pressure is high it causes the box to bend inwards;
this movement is transferred by levers, causing a
needle to turn on the scale. When atmospheric
pressure is low, the pressure inside the box is
greater than that outside the box, and the box
bends outwards.

altimeter (*n*) an aneroid barometer (↑) specially
calibrated to measure height, by measuring the
decrease in atmospheric pressure. As altitude
(height from the Earth's surface) increases,
atmospheric pressure decreases.

suction (*n*) the process of drawing air out of a space and making a partial vacuum (p.31). A fluid (air or liquid) then fills the space; the fluid is moved into the space by atmospheric pressure (p.31), which is greater than the pressure of the partial vacuum (p.31) caused by the suction.

syringe (*n*) a device used for sucking in liquids and forcing them out again under pressure. When the plunger in the barrel of the syringe is raised, atmospheric pressure pushes liquid through the nozzle. When the plunger is lowered, the liquid is forced out.

siphon (*n*) a device which makes use of atmospheric pressure to draw a liquid out of a vessel, along a tube, and pass it to a lower level. It consists of an inverted U-shaped tube with one arm longer than the other. The end of tube D must be lower than the end B in order to siphon liquid from vessel A down to vessel C.

manometer (*n*) an instrument for measuring gas pressure. It usually consists of a U-tube containing liquid. Gas is fed into the tube; the pressure of the gas pushes the liquid round the tube, and is measured from the difference in the levels of the liquid in the two arms.

valve[1] (*n*) (1) a device that controls the flow of fluids through an opening or along a pipe. (2) any mechanical device that allows fluids to flow in one direction only.

pump (*n*) a device used to raise or move water or other liquids, or to compress gases, i.e. to force them into smaller spaces.

common pump a device used to raise water or other liquids. A piston moves up and down in the barrel of the pump. Liquid is forced into the barrel by atmospheric pressure and liquid is pushed out through an outlet. This pump is unable to raise water more than 10 metres.

lift pump an alternative name for **common pump**. *see* **force pump**.

force pump a pump that works like a common pump (↑) to draw liquid into its barrel, but forces the liquid out under pressure. This is done by a force pushing down a piston that in turn pushes out the water. A force pump can raise liquids to a height greater than 10 metres.

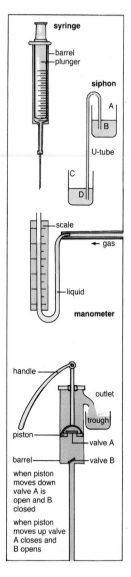

syringe
— barrel
— plunger

siphon
A
B
U-tube
C
D

scale
← gas
liquid
manometer

handle
outlet
trough
piston
valve A
barrel
valve B

when piston moves down valve A is open and B closed

when piston moves up valve A closes and B opens

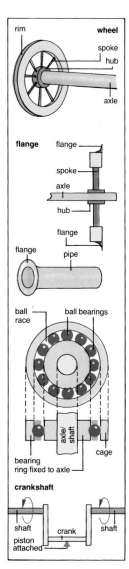

techology (*n*) the study of science as used in industry, industrial processes and in engineering.

mechanics (*n*) the study of the forces acting on solid bodies and the motion that can be produced by such forces. It includes the subjects of the properties of materials; force, work and energy; machines and engines; hydraulics; and motion of solids and liquids. **mechanical** (*adj*).

wheel (*n*) a circular object connected at the centre to an axle or a shaft (↓). The wheel may be fixed to the axle or may be free to move round the axle. It can be solid or have an outer frame joined to an inner part, the hub (↓), by spokes. A wheel can be used for making vehicles move, or in machinery.

hub (*n*) the inner part of a wheel, connected to the axle, or to a shaft (↓), or moving round an axle.

flange (*n*) a raised edge that stands out from the main part of a body, e.g. the flange on a wheel; a flange on a pipe. The flange on a wheel keeps the wheel on a railway line.

shaft (*n*) (1) a long, straight, round rod of metal used for the transmission of power from one place to another. A shaft is usually driven by a wheel, or gear wheel (p.34) and usually drives another wheel; it transmits torque (p.19). (2) a hole going straight down into a mine.

bearing (*n*) a support for a rotating shaft (↑) to keep it in position, or a part of a machine which turns on a stationary rod or shaft. Frictional resistance between a bearing and a shaft produces heat, so lubrication is needed to reduce friction.

ball bearing (1) a ring of hard steel balls between a shaft (↑) and a bearing (↑), either of which rotates. The steel balls are carried in a metal cage, at a suitable distance apart, evenly distributed round the bearing. Frictional resistance between the balls and the cage is very low, and little lubrication is needed for free running of the bearing, only sufficient to prevent rusting. (2) a single ball in a ball bearing.

ball race the ring containing the steel ball bearings.

crankshaft (*n*) a shaft with an arm at right angles. A reciprocating motion of a piston going up and down is converted to a rotary motion of the shaft. The arm at right angles is a crank. A reciprocating steam engine and an internal combustion engine use a crankshaft to provide a torque (p.19).

gear (*n*) a system of moving parts, which are usually wheels, used in the transmission (p.18) of power (p.13), to increase or decrease speeds of rotation, or to change the direction of rotation. The wheels used in gears are mainly gear wheels (↓), although friction (p.14) drives, chain drives, and belt drives may be used.

gear wheel a wheel which has teeth around its circumference (p.11). The teeth engage with the teeth of another wheel and drive it round. Gear wheels are a very important method of transmission of power and motion. The velocity ratio of two gear wheels determines the relative rate of rotation of each:

$$\frac{\text{revolutions per minute of wheel A}}{\text{revolutions per minute of wheel B}}$$
$$= \frac{\text{number of teeth in wheel B}}{\text{number of teeth in wheel A}}$$

gearbox (*n*) the part of a motor car which contains a set of gear wheels (↑). The gear wheels are so arranged that for a given engine speed, the road wheels can be rotated slower or quicker to give greater or less power. Gearboxes usually contain four forward gears, and one reverse, although some gearboxes contain three forward and one reverse.

differential (*n*) a type of gear used on a motor car to allow the back wheels to turn at different speeds when the car is going round a curve. The inner wheel is slowed down, and the outer wheel speeded up; the average speed of the two wheels is that of the crown wheel, driven by the transmission shaft.

pinion (*n*) a small gear wheel (↑) driving, or driven by, a larger gear wheel (↑).

chassis (*n*) the metal frame which supports the body and the engine of a motor car; it includes the road wheels and transmission (p.18).

shock absorber a device which slows down the action of the suspension springs of a vehicle. It consists of a piston operating in a cylinder containing oil. The piston forces the oil out of the cylinder through a small hole, slowing the bounce of the spring; the oil flows back into the cylinder through another hole when the spring recovers from the bounce.

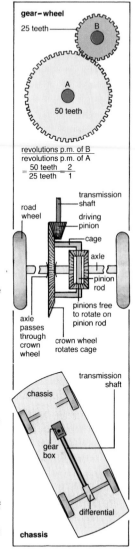

gear–wheel
25 teeth
B
A
50 teeth

$$\frac{\text{revolutions p.m. of B}}{\text{revolutions p.m. of A}}$$
$$= \frac{50 \text{ teeth}}{25 \text{ teeth}} = \frac{2}{1}$$

transmission shaft
road wheel
driving pinion
cage
axle
pinion rod
pinions free to rotate on pinion rod
axle passes through crown wheel
crown wheel rotates cage

transmission shaft
chassis
gear box
differential
chassis

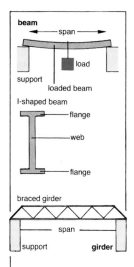

beam (*n*) a long structure, supported either at both ends, or at one end only, with forces acting at right angles to its length. A beam can be a square piece of wood, or a piece of steel shaped like a letter U or a letter I. When a load is placed on a beam, one surface is stretched by tension and the opposite surface is shortened by compression.

girder (*n*) a complex beam (↑) formed from rods or pieces of steel bolted or riveted together. The pieces of steel joining the long sections of steel are braces or trusses.

span (*n*) the distance between two limits of space or time; the distance between the supports of a beam (↑) or girder (↑).

cantilever (*n*) a beam, or girder, supported at one end and free at the other end.

derrick (*n*) a machine for lifting heavy loads, usually found on ships. A mast, or support, has a beam (↑) hinged at the bottom of the mast. The beam is supported by a rope passing over a pulley at the end of the beam, and lifts a load by power supplied by a winch (↓).

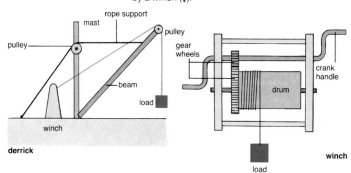

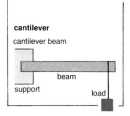

winch (*n*) a device for raising heavy loads, consisting of a drum driven by a pair of gear wheels which are operated by a crank handle. The handle gives a mechanical advantage (p.15) to the first gear wheel, and this is increased by the second gear wheel. **winch** (*v*).

hoist (*n*) a single pulley used to raise a load by pulling downwards on a rope. *See* **tension** (p.16). **hoist** (*v*).

machine tool a tool operated by a motor, used for cutting or stamping metal objects to a required shape.

lathe (*n*) a machine for cutting wood or metal into round shapes. The material is rotated at a fast speed, and a sharp tool held against a support to cut the material to the required shape. A lathe is also used for making holes in wood or metal, for cutting screws or bolts, and for polishing wooden or metal objects.

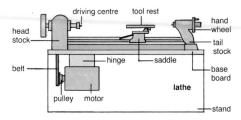

lathe

jig (*n*) a device for keeping wood or metal in position, and for guiding a tool to cut, shape or drill as required. A jig is used with machine tools (↑) and ensures accurate measurements for any operation.

weld (*v*) to join two pieces of metal, usually iron or steel, by heating them and then hammering them together so that they form one piece of metal. Also to join the pieces by melting them using an oxyacetylene flame; on cooling they are joined as one piece.

thin metal rivets have head flattened by mechanical pressure

rivet

rivet (*n*) a type of nail used to join two sheets of metal. The rivet is heated red-hot, pushed through holes in the plates and hammered flat. On cooling, the rivet contracts and fastens the two sheets permanently together.

forge (*v*) to hammer hot pieces of metal, usually steel or soft iron, into a required shape. The metal is hammered by hand or pressed by machine.

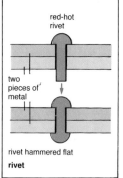

two pieces of metal

red-hot rivet

rivet hammered flat

rivet

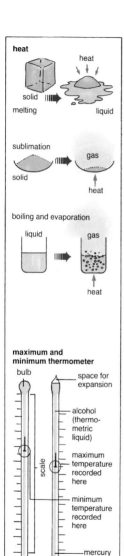

heat

heat

solid

melting   liquid

sublimation

gas

solid

heat

boiling and evaporation

liquid   gas

heat

maximum and minimum thermometer

bulb

space for expansion

alcohol (thermo-metric liquid)

maximum temperature recorded here

scale

minimum temperature recorded here

mercury

heat (*n*) a form of energy which substances possess because of the kinetic energy (p.14) of their molecules. Heat is produced by the accelerated vibration of molecules. It is measured in Joules, J. The physical effects of a change in heat are (1) a change in temperature; (2) a change of state from solid to liquid (melting), solid to gas (sublimation) or liquid to gas (boiling and evaporation); (3) expansion or contraction. Heat passes from a higher temperature substance to a lower temperature substance. **heat** (*v*).

temperature (*n*) a measure, on a scale, used for comparing the degree of hotness or coldness of something.

thermometer (*n*) an instrument for measuring temperature. It uses a material that expands noticeably with a small change in temperature. Most thermometers are made of a thin glass tube containing mercury or alcohol. As the temperature increases, the liquid expands and rises up the tube. The temperature can be read off the graduated scale.

maximum and minimum thermometer a thermometer (↑) designed to measure the highest (maximum) and lowest (minimum) temperatures (↑) recorded over a length of time, usually one day. When the temperature rises, the alcohol expands pushing the fine mercury thread along the right-hand tube. A steel index marks the highest temperature reached. The index remains in place after the temperature has fallen. When the temperature falls, the alcohol contracts and the level of mercury in the left-hand tube rises, and a steel index records the minimum temperature. The indices are reset by a magnet.

fixed point (1) a known temperature that can be accurately reproduced, e.g. the boiling point of a liquid. (2) either of two reference points of a thermometer scale; the **upper fixed point** is the temperature of boiling water; the **lower fixed point** is the temperature of melting ice.

degree (*n*) a unit on a scale or on a measuring instrument, e.g. a right angle is divided into 90 equal units, called degrees; the scale of temperature is divided into a number of units between two fixed points (↑).

Celsius scale the centigrade thermometer scale. The symbol is °C. The lower fixed point (0°C) is the temperature of melting ice, the upper fixed point (100°C) is the temperature of boiling water.

Fahrenheit scale a temperature scale with the lower fixed point (p.37) at 32 degrees and the upper fixed point at 212 degrees. The symbol is °F.

kelvin (n) the SI unit of temperature; the symbol is K. 1 K = 1°C. The Kelvin scale has the lower fixed point at 273 K and the upper fixed point at 373 K.

expand (v) of a solid, to increase in length, area or volume; of a fluid, to increase in volume. The increase is due to a rise in temperature, although gases may also expand due to a decrease in pressure. *See* **Boyle's law** (p.29). **expansion** (n) the increase in length, area or volume.

apparent expansion the expansion of a fluid relative to the vessel in which it is contained. On heating, both the fluid and the vessel expand, so the true expansion of the fluid is its apparent expansion plus the expansion of the vessel.

contract (v) of a solid, to decrease in length, area or volume; of a fluid, to decrease in volume. A fall in temperature causes the decrease, although gases may contract due to an increase in pressure. *See* **Boyle's law** (p.29). **contraction** (n).

coefficient (n) a constant ratio used to measure changes in a particular quantity of a material with respect to a specific property, e.g. the coefficient of expansion is the increase in length, area or volume per unit length, area or volume for a rise in temperature of one degree.

bimetallic (adj) made of two different metals fastened together. a bimetallic strip is made of two metals with different rates of expansion when heated. The unequal expansion results in the strip bending. A bimetallic strip can be used for measuring temperature, e.g. a bimetallic thermometer, or for regulating temperature, e.g. a bimetallic strip in a thermostat (↓).

thermostat (n) a device for regulating temperature. It uses a bimetallic (↑) strip that bends when hot and straightens when cold. When cold it makes an electric contact to start a heating coil. When the temperature rises above a determined set value, the contact is broken, and the temperature falls.

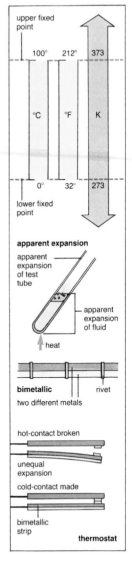

upper fixed point

	100°	212°	373
	°C	°F	K
	0°	32°	273

lower fixed point

apparent expansion

apparent expansion of test tube

apparent expansion of fluid

heat

bimetallic   rivet

two different metals

hot-contact broken

unequal expansion

cold-contact made

bimetallic strip

thermostat

transfer (*v*) to move matter or energy from one position to another, e.g. to transfer heat from a hot solid to a cold liquid. Heat is transferred by convection, conduction or radiation. **transfer** (*n*).

conduction (*n*) the movement of heat energy through a solid object. Fluids do not readily conduct heat. Also, the movement of an electric current through a substance. **conduct** (*v*), **conductor** (*n*).

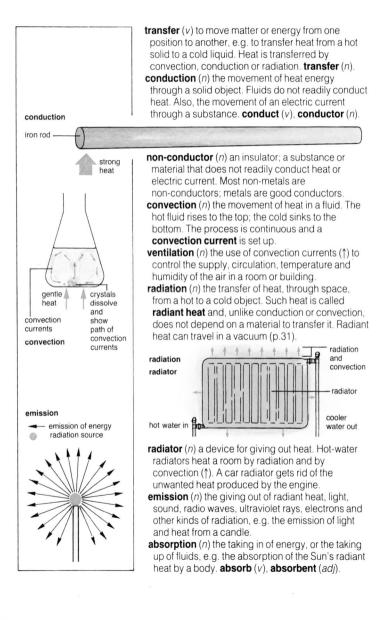

conduction

iron rod

strong heat

gentle heat

convection currents

crystals dissolve and show path of convection currents

convection

emission

◀— emission of energy

● radiation source

non-conductor (*n*) an insulator; a substance or material that does not readily conduct heat or electric current. Most non-metals are non-conductors; metals are good conductors.

convection (*n*) the movement of heat in a fluid. The hot fluid rises to the top; the cold sinks to the bottom. The process is continuous and a **convection current** is set up.

ventilation (*n*) the use of convection currents (↑) to control the supply, circulation, temperature and humidity of the air in a room or building.

radiation (*n*) the transfer of heat, through space, from a hot to a cold object. Such heat is called **radiant heat** and, unlike conduction or convection, does not depend on a material to transfer it. Radiant heat can travel in a vacuum (p.31).

radiation

radiator

radiation and convection

radiator

hot water in

cooler water out

radiator (*n*) a device for giving out heat. Hot-water radiators heat a room by radiation and by convection (↑). A car radiator gets rid of the unwanted heat produced by the engine.

emission (*n*) the giving out of radiant heat, light, sound, radio waves, ultraviolet rays, electrons and other kinds of radiation, e.g. the emission of light and heat from a candle.

absorption (*n*) the taking in of energy, or the taking up of fluids, e.g. the absorption of the Sun's radiant heat by a body. **absorb** (*v*), **absorbent** (*adj*).

humidity (*n*) the amount of water vapour in the air. When the air is full of water vapour we say it is saturated; when it has no water vapour it is completely dry air. The **relative humidity** of air is the ratio of the water vapour in the atmosphere at a given temperature to the maximum or saturated vapour content at the same temperature. It is expressed as a percentage — 100% is fully saturated air; 0% is completely dry air.

hygrometer (*n*) an instrument used to measure the humidity of the atmosphere, e.g. a hair hygrometer measures humidity by means of a single human hair that expands with increasing humidity and contracts with decreasing humidity.

wet-and-dry bulb thermometer an instrument for determining the relative humidity (↑) of the atmosphere. It consists of two identical thermometers. The wet-bulb thermometer is kept wet and records the lowering of temperature caused by the evaporation of the water (the evaporation depends on the relative humidity of the air). The dry-bulb thermometer records atmospheric temperature. The difference in the two readings is a measure of the humidity.

dew (*n*) small drops of water that are formed on surfaces when warm air cools (condenses).

dew point the temperature at which dew forms when the atmosphere becomes saturated with water vapour and condensation occurs.

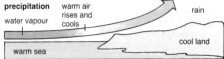

precipitation   warm air rises and cools

water vapour

rain

cool land

warm sea

precipitation (*n*) rainfall, snow showers, hail, sleet, fog and mist. Precipitation results from the condensation of water vapour into water droplets as the temperature lowers. When the temperature falls below the dew point, the droplets fall as rain. If the temperature falls below freezing point the water vapour forms ice crystals, which fall as snow.

mist (*n*) very small drops of water formed by condensation (↓) of water vapour when warm, moist air is suddenly cooled. Mist forms at ground level. Fog forms when cooling water vapour condenses on dust in the air.

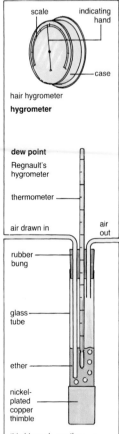

scale   indicating hand

case

hair hygrometer
hygrometer

dew point
Regnault's hygrometer

thermometer

air drawn in

air out

rubber bung

glass tube

ether

nickel-plated copper thimble

thimble cools as ether evaporates in stream of air bubbles. The temperature is noted when condensation appears. Bubbling is stopped and thimble temperature rises until no dew is seen. The mean of the two temperatures is the dew point.

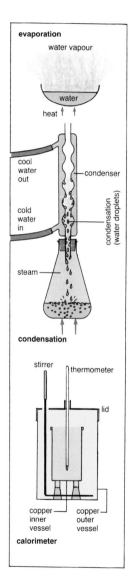

evaporation

water vapour

water

heat ↑ ↑

cool water out

condenser

cold water in

condensation (water droplets)

steam

condensation

stirrer

thermometer

lid

copper inner vessel

copper outer vessel

calorimeter

evaporation (*n*) the changing of a liquid into a vapour by heat or moving air, e.g. the evaporation of water into water vapour at, or below, its boiling point.

condensation (*n*) the changing of a vapour or gas into a liquid by cooling, e.g. the formation of water droplets when steam meets a cold surface. Compare **liquefaction**, the changing of a vapour to a liquid just by increasing the pressure.

calorie (*n*) a former unit of measurement of heat. One calorie is the quantity of heat needed to raise the temperature of one gram of water by 1°C. 1 calorie = 4.18 J.

calorimeter (*n*) an apparatus for measuring quantities of heat, usually by observing the rise in temperature of a known mass of water.

heat capacity the number of joules required to increase the temperature of an object by 1 K (1°C). A change in the heat content of an object is measured by its heat capacity multiplied by the change in temperature, e.g. the heat capacity of a copper vessel is 40 J/K; 360 J are needed to raise its temperature by 9 K.

specific heat capacity the number of joules required to raise the temperature of 1 kg of a substance by 1 K (1°C), e.g. the specific heat capacity of water is 4180 J per kg per kelvin.

latent heat the heat (given out or taken in) that changes the physical state of a substance without changing its temperature, e.g. the heat required to change water at 100°C to water vapour at 100°C.

specific latent heat the number of joules required to change 1 kg of a substance from solid to liquid, or liquid to gas, without changing the temperature. Each substance possesses two specific latent heats: (1) of melting; (2) of vaporization. These two quantities are constant for a specific substance.

refrigeration (*n*) the use of energy to remove heat from an object. In a refrigeration system, a liquid, such as ammonia, is pumped through pipes in the cold chamber of a refrigerator. Here it absorbs heat from the contents and becomes a vapour. This is compressed by a pump, and becomes a liquid again. By constant circulation of the substance, heat is taken away from the inside of the refrigerator and passed to the outside air.

states of matter the physical condition of matter;
either solid, liquid or gaseous. These are the three
states of matter.

solid (*n*) a solid has a definite shape and a definite
volume. It is formed from molecules held together
by powerful forces. **solid** (*adj*), **solidify** (*v*).

liquid (*n*) a liquid has no fixed shape, it will take that
of the container in which it is held, but it has a
definite volume. The forces between the molecules
in a liquid are weaker than those between the
molecules in a solid. **liquid** (*adj*), **liquefy** (*v*).

gas (*n*) a gas has no definite volume and no definite
shape. The molecules in a gas are free to move
about, with no forces holding them together.
gaseous (*adj*).

fluid (*n*) a substance that is able to flow, i.e. a liquid
or a gas. **fluid** (*adj*).

Charles' law at a constant temperature (p.37), the
volume (p.11) of a fixed mass of gas (↑) is
proportional to its temperature (measured in kelvin).
In symbols: $V \propto T$.

gas law a combination of Boyle's law (p.29) and
Charles' law. It states that for a fixed mass of gas,
(pressure) × (volume) is proportional to
(temperature); the temperature is measured in
kelvin. In symbols: $pV \propto T$.

change of state the change of a substance from one
of the physical states of matter (solid, liquid and
gaseous) into another.

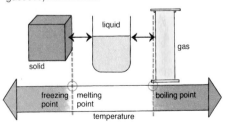

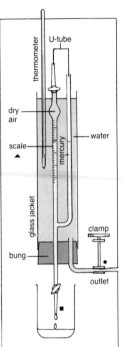

Charles' law
apparatus

water run out at ● and jacket
filled with hot water

■ mercury run out to equal
levels in U-tube

▲ scale measures volume of
dry air at various
temperatures

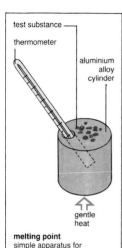

test substance

thermometer

aluminium alloy cylinder

gentle heat

melting point
simple apparatus for
temperatures up to 250°C

melt (*v*) the change from a solid to a liquid occuring
when the solid is heated at an appropriate constant
temperature, the melting point (↓). Compare
dissolve.

melting point the temperature at which a solid
changes to a liquid. Each solid has a particular
melting point, e.g. the melting point of ice is 0°C, the
melting point of iron is 1539°C. It is a physical
property that may be used to identify a substance.

substance	melting point	substance	m.p°C
benzene	0.5°C	ethanedioic acid	190
glucose	142	sodium chloride	804
napthalene	80	zinc chloride	262
potassium	254	iron (II) chloride	670
lead	327	calcium oxide	2572
magnesium	650	calcium chloride	772

solidify (*v*) to change from a liquid to a solid, by
cooling, e.g. fats solidify on cooling. Compare
crystallization (p.48) and **sublimation** (p.106).

freezing point the temperature at which a liquid
changes to a solid, by cooling. The freezing point
and the melting point are the same temperature for
a pure substance.

boil (*v*) to change from a liquid to a vapour (↓), by
heating at a constant temperature, e.g. water boils
at 100°C to form steam.

boiling point the temperature at which a liquid
becomes a vapour. Each liquid has a particular
boiling point, e.g. the boiling point of water is 100°C,
the boiling point of ethanol is 78°C at normal
atmospheric pressure. It is a physical property that
may be used to identify a substance.

substance	boiling point	substance	b.p.°C
benzene	80°C	ethanoic acid	118
napthalene	218	trichloromethane	61
iron	2900	aminobenzene	184
phosphorus	280	lead (II) iodide	954
potassium	960	lead	1750
magnesium	1100	glycerol	290

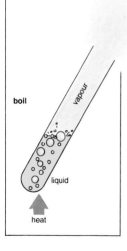

boil

vapour

liquid

heat

vapour (*n*) the gaseous form of a substance that is
usually a liquid or solid at ordinary temperatures. A
vapour can be changed to a liquid by increasing
the external pressure without lowering the temperature; a gas cannot be liquefied by pressure alone.

element (*n*) a simple substance of one type of atom, all with the same atomic number (p.65). Elements cannot be decomposed into simpler substances by chemical change.

allotropes (*n.sing. allotrope*) different physical forms of the same element, e.g. diamond, charcoal, graphite and coal are allotropes of carbon. Allotropes of an element have the same chemical properties. **allotropic** (*adj*).

metal (*n*) an element characterized by the properties of lustre (↓), ductility (↓), malleability (↓), and conductance of heat and electricity. In chemical reactions the atoms of metals form positive ions. Some examples of metals are copper and zinc. **metallic** (*adj*).

lustre (*n*) the quality or condition of a surface of shining by reflected light. Silver, gold and other metals have lustre as do gems and precious stones.

ductile (*adj*) describes a substance that is capable of being drawn out into a wire. **ductility** (*n*).

malleable (*adj*) describes a substance that can be beaten into shapes. **malleability** (*n*).

brass — 70% copper, 30% zinc

Delta metal — 55% copper, 40% zinc, 5% (Al,Fe,Mn)

German silver — 50% copper, 30% zinc, 20% nickel

bronze — 92% copper, 8% tin

alloy (*n*) a metallic material formed when two or more metals are melted together, or when a metal and non-metal are combined. Brass, bronze and steel are alloys.

non-metal (*n*) an element that does not possess the general properties of a metal. Non-metals are usually poor conductors, e.g. oxygen and nitrogen. They may be solid, liquid or gaseous.

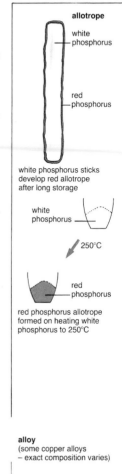

allotrope

white phosphorus

red phosphorus

white phosphorus sticks develop red allotrope after long storage

white phosphorus

250°C

red phosphorus

red phosphorus allotrope formed on heating white phosphorus to 250°C

alloy
(some copper alloys – exact composition varies)

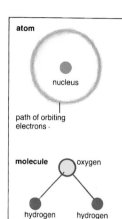

atom

nucleus

path of orbiting electrons ·

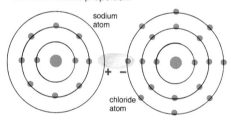

molecule

oxygen

hydrogen   hydrogen

e.g. water molecule

substance
e.g. salt
(sodium
chloride)

sodium atom

chloride atom

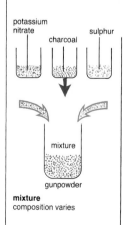

potassium nitrate

charcoal

sulphur

mixture

gunpowder

mixture
composition varies

atom (*n*) the smallest particle of any chemical element that possesses all the properties of that element, and can enter into chemical reactions. Each atom has a central nucleus with one or more orbiting electrons.

molecule (*n*) the smallest particle of an element or compound that can exist by itself and still retain all the properties of that element or compound. A molecule of an element consists of one or more atoms of that element; a molecule of a compound consists of one or more atoms of each element of that compound. **molecular** (*adj*).

compound (*n*) a substance that contains two or more chemical elements in constant proportion by mass. The elements are held together by chemical bonds and cannot be separated by physical means. A compound can be given an exact chemical formula (p.63).

composition (*n*) the elements forming a compound, and their relative proportion.

substance (*n*) a chemical element or compound with a definite composition. A substance has specific characteristic properties by which it can be identified, e.g. salt, calcium, starch.

decompose (*v*) to breakdown a chemical substance into simpler substances by chemical action, heat, reduced pressure or an electric current.

material (*n*) something that can be recognized and named. Its properties and chemical composition are variable within limits, but allow identification.

mixture (*n*) a material formed by mixing together two or more substances in any proportion; a chemical reaction does not take place. The substances can be separated again by physical means.

constituent (*n*) a single substance or element that forms part of a compound or mixture.

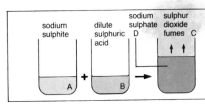

sodium sulphite — dilute sulphuric acid — sodium sulphate D — sulphur dioxide fumes C

A + B →

chemical reaction
substance A added to
substance B forms
substances C and D

C and D are products
product

chemical reaction any chemical change that occurs when two or more substances are put together. The reaction produces a new substance or substances.

product (*n*) a substance or material produced by a chemical reaction or process.

reversible reaction a chemical reaction which can occur in two directions, depending on conditions. The products of a reversible reaction react chemically with each other to produce the original substances.

chain reaction

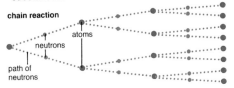

atoms

neutrons

path of neutrons

chain reaction a chemical reaction in which a substance, A, produces a product, B, and B then produces more of substance A. If each molecule of A generates one new molecule of A, the reaction is maintained. If each molecule of A produces more than one new molecule of substance A, the reaction is explosive, e.g. one neutron and one atom of uranium-235 produce 2 neutrons and other products.

replaceable (*adj*) in a chemical reaction, describes an element that can be displaced, and its place taken by a different substance.

dehydration (*n*) the process of removing water from a substance by heat or by a chemical reaction, e.g. milk can be dehydrated to form powdered milk. **dehydrate** (*v*).

effervesce (*v*) of a liquid, to produce gas bubbles rapidly due to chemical reaction, e.g. when zinc is placed in warm dilute sulphuric acid, the liquid effervesces. **effervescence** (*n*), **effervescent** (*adj*).

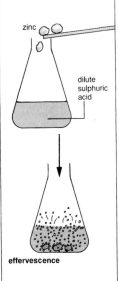

zinc

dilute sulphuric acid

effervescence

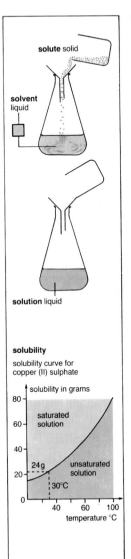

solute solid

solvent liquid

solution liquid

solubility

solubility curve for copper (II) sulphate

solubility in grams

saturated solution

unsaturated solution

24 g

30°C

temperature °C

dissolve (v) to make a solid substance or gas disappear into a liquid by mixing. Compare **melt**.

solute (n) any substance, solid or gas, that will dissolve when added to water or other liquid, e.g. sulphur dioxide is a solute in water.

solvent (n) the liquid in which a solute (↑) dissolves, e.g. water is the solvent and sugar the solute in a sugar solution.

solution (n) a liquid containing a solute dissolved in a solvent (↑), e.g. a salt solution consists of common salt (solute) dissolved in water (solvent). Unless another solvent has been specified it is assumed that the solvent is water.

saturated (adj) unable to absorb, or contain, more. A saturated gas is one that cannot contain any more water vapour; a saturated solution is one that is unable to dissolve any more solute (↑); a saturated solid is one that is unable to absorb any more liquid. **saturate** (v).

unsaturated (adj) able to absorb, or contain, more. An unsaturated gas can contain more water vapour; an unsaturated solution can contain more solute.

solubility (n) the mass, in grams, of a solute (↑) that will dissolve in 100 grams of a solvent (↑) at a given temperature to form a saturated solution, e.g. the solubility of copper (II) sulphate is 24 g at 30°C, so 24 g of the solute dissolved in 100 g of water at 30°C forms a saturated solution. **soluble** (adj).

insoluble (adj) describes a solid or a gas that will not dissolve in a given liquid. Some substances that are insoluble in water are soluble in alcohol or other liquids, e.g. some fatty substances are soluble in alcohol but not in water.

concentration (n) (1) a measure of the amount of a solute dissolved in a given volume of its solution. If the relative amount of the solute is large, the concentration of the solution will be high. It can be expressed as (a) grams of solute per litre of solution $(g\,dm^{-3})$; (b) moles (p.00) per dm^3; (c) a percentage by mass. (2) the process of concentrating, i.e. increasing the concentration of a solution by evaporating the solvent. **concentrate** (v), **concentrated** (adj).

dilute (v) to reduce the concentration of a solution by increasing the quantity of the solvent, e.g. to dilute concentrated acid by adding water (solvent).

crystal (*n*) a clear, solid substance with a regular geometric shape. The crystal form is due to the regular pattern of the particles in the substance. Crystals of one substance are all the same shape; different substances have crystals of different shapes. **crystalline** (*adj*) a crystalline solid is one that is composed of crystals. **crystallize** (*v*).

crystallization (*n*) the process of forming crystals by cooling (evaporating) the solvent in which the crystals are dissolved.

water of crystallization a definite proportion of water molecules that are combined with ions in the structure of crystals when the crystals are formed from a solution in water. Removing the water of crystallization will alter the structure of the crystal, e.g. copper (II) sulphate forms blue crystals; when its water of crystallization is removed it becomes a white powder.

anhydrous (*adj*) describes either crystals having no water of crystallization, or dry amorphous (↓) solids.

hydrated (*adj*) describes crystals containing water of crystallization (↑).

tetrahedral (*adj*) describes an object with a shape like a tetrahedron, i.e. a solid figure with four flat triangular faces.

amorphous (*adj*) describes a solid that does not have a crystalline structure; it is without definite form, e.g. rubber.

suspension (*n*) tiny insoluble particles evenly spread throughout a liquid or gas. If allowed to stand, the particles will settle very slowly to the bottom of the container and form a sediment.

colloid (*n*) one of many particles that do not dissolve, but remain permanently suspended in a fluid. They are larger than molecules in solution, but smaller than molecules in suspension. The particles of fat suspended in milk are colloids; milk is a colloidal solution. **colloidal** (*adj*).

emulsion (*n*) a colloidal solution formed from two liquids; one liquid is suspended in the other, e.g. oil and water when shaken form an emulsion. If the emulsion is not fixed by a special substance (emulsifying agent) the liquids will separate into layers.

aerosol (*n*) a mist of fine particles of a liquid dispersed in air.

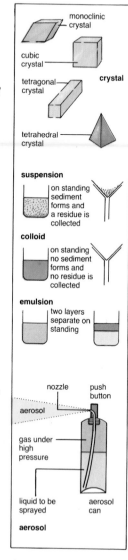

monoclinic crystal

cubic crystal

tetragonal crystal

crystal

tetrahedral crystal

suspension

on standing sediment forms and a residue is collected

colloid

on standing no sediment forms and no residue is collected

emulsion

two layers separate on standing

nozzle　push button

aerosol

gas under high pressure

liquid to be sprayed

aerosol can

aerosol

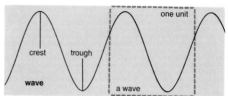

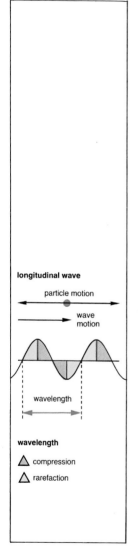

wave (*n*) one unit of a series of troughs and crests as formed on the surface of water when set in motion by a vibrating rod.

wave motion the transfer of energy in the form of a wave (↑). The particles in a material medium do not move forward; they move about a central position and the disturbance, represented by waves, moves forward.

transverse wave a wave in which the disturbance of the particles of a material medium is at right angles to the direction of the wave.

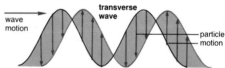

longitudinal wave a wave in which the disturbance of the particles of a material medium is backwards and forwards (about a central position) in the direction of the wave.

wavelength (*n*) the distance between the crest of one wave and the crest of the next wave moving in the same direction, i.e. the distance between a point on one wave and the corresponding point on the next wave moving in the same direction. The symbol for wavelength is λ. The wavelength λ is equal to the speed of the wave motion (*v*) divided by its frequency (*f*).

frequency[1] (*n*) the number of vibrations (usually per second) in a wave motion. In a transverse wave this is taken to be the number of crests (or troughs) that pass a fixed point in 1 second. The symbol for frequency is *f*; frequency is measured in Hertz (Hz).

amplitude (*n*) (1) the distance between the top (or bottom) and the middle of a wave. (2) the distance between the central and outer positions occupied by a vibrating body.

stationary wave the wave formed when two identical
waves travelling simultaneously in opposite
directions meet, e.g. the wave seen when a
stretched string vibrates. A stationary wave does
not move along a medium. The wavelength of a
stationary wave is the distance between a point on
one wave and the corresponding point on the next
but one wave, i.e. twice the distance between two
nodes (↓).

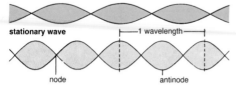

stationary wave

├─── 1 wavelength ───┤

node   antinode

node[1] (*n*) a point on a stationary wave where there is
no amplitude (p.49) of vibration.

antinode (*n*) the point of maximum amplitude (p.49)
of vibration in a stationary wave.

fundamental frequency the lowest possible
frequency of vibration of a stationary wave; the
lowest frequency of a musical note. The
fundamental frequency of a stretched string is
produced when the wavelength is half the length of
the string. A musical instrument produces a **pure
note** when it vibrates with its fundamental
frequency alone. A musical note usually consists of
a fundamental frequency and several overtones (↓)
of higher frequency. The overtones give a residual
note its characteristic quality.

overtone (*n*) a note of higher frequency than the
fundamental frequency (↑) given out by a musical
instrument.

natural frequency the frequency of vibration of a
solid object, or a column of fluid, when free to
vibrate and not acted upon by any external force.

resonance (*n*) the creation of very large amplitudes
of vibration in an object, produced when a periodic
force with the same frequency as the natural
frequency (p.49) of an object is applied to the
object, e.g. a note of exactly the same frequency as
the natural frequency of a wine glass can cause it to
vibrate violently and even break. The violent
vibration is resonance.

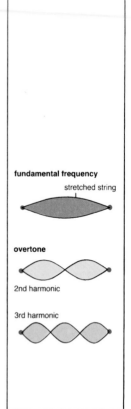

fundamental frequency

stretched string

overtone

2nd harmonic

3rd harmonic

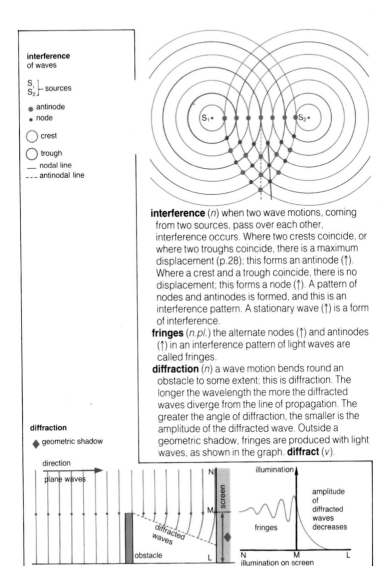

interference
of waves

S_1 ⎤
S_2 ⎦ ⟩ sources

● antinode
· node

◯ crest

◯ trough

— nodal line
--- antinodal line

diffraction

◆ geometric shadow

interference (*n*) when two wave motions, coming from two sources, pass over each other, interference occurs. Where two crests coincide, or where two troughs coincide, there is a maximum displacement (p.28); this forms an antinode (↑). Where a crest and a trough coincide, there is no displacement; this forms a node (↑). A pattern of nodes and antinodes is formed, and this is an interference pattern. A stationary wave (↑) is a form of interference.

fringes (*n.pl.*) the alternate nodes (↑) and antinodes (↑) in an interference pattern of light waves are called fringes.

diffraction (*n*) a wave motion bends round an obstacle to some extent; this is diffraction. The longer the wavelength the more the diffracted waves diverge from the line of propagation. The greater the angle of diffraction, the smaller is the amplitude of the diffracted wave. Outside a geometric shadow, fringes are produced with light waves, as shown in the graph. **diffract** (*v*).

acoustics (*n*) (1) the study of sound, its production, propagation, and the effects it causes. (2) the effect of the structure and materials of a building on the ability to hear sounds produced in the building, e.g. the acoustics of a theatre. **acoustic** (*adj*).

vibration (*n*) the rapid backwards and forwards movement of an elastic solid, or particles of a fluid, when displaced from their position of rest. e.g. the motion of a stretched string when it is plucked. **vibrate** (*v*). **vibrator** (*n*). **vibratory** (*adj*).

frequency² (*n*) the number of times an event occurs within a given time, e.g. the frequency of vibrations of an object is the number of vibrations in 1 second.

sound wave a longitudinal pressure wave, possessing a speed of about 330 m/s in air, produced by a vibrating object. Sound can only travel through a material medium, i.e. a gas, liquid, or solid. The vibrating object emits: (1) a wave of high pressure, known as a **compression**; (2) a wave of low pressure, a **rarefaction**. Compressions and rarefactions follow each other alternately. In air, the pressure between the two waves returns to atmospheric pressure. The distance between one compression and the next is the wavelength of the sound wave.

decibel (*n*) the practical unit for measuring the loudness of a sound; it describes how many times one sound is more intense than another. Two intensities of sound, P_1 and P_2, differ by n decibels where; $n = 10 \log_{10} P_2/P_1$. The intensity of a sound, P_2, under observation, is compared to a sound of intensity, P_1, as a reference level; this is usually the intensity of the lowest sound of the same frequency that can be heard by the human ear.

ultrasonic (*adj*) describes a sound wave with a frequency too great to be heard by human beings; generally taken as a sound wave with a frequency greater than 20000 Hz. Contrast **supersonic** (p.131).

period (*n*) the time taken for a regularly occuring event to make one complete cycle.

periodic (*adj*) describes an event that occurs at regular intervals, e.g. the periodic appearance of Halley's Comet (every 75 years).

hertz (*n*) the SI unit of frequency. The symbol is Hz; 1 hertz is equal to 1 cycle per second. An event has a frequency of 1 Hz if its period is 1 second.

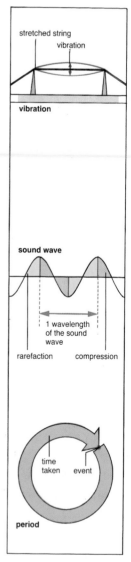

stretched string

vibration

vibration

sound wave

1 wavelength of the sound wave

rarefaction   compression

time taken   event

period

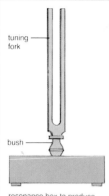

tuning fork

bush

resonance box to produce a strong sustained note

musical note

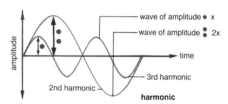

harmonic

musical note a sound arising from an object that produces regular and repeating vibrations of the air. A musical note depends on a source to produce the regular vibrations, a medium in which the sound waves (↑) can travel, and reception by the hearer.

pitch 1 (*n*) the highness or lowness of a sound that determines its position in a musical scale. Pitch depends on the frequency of the sound waves produced by the source of the note. The higher the frequency the higher the note.

quality (*n*) that characteristic of a musical note that allows it to be distinguished from notes of the same pitch and loudness made by different instruments, e.g. notes produced by a clarinet and a saxophone are easily distinguished because of their different qualities. The quality or **timbre** of a note depends on the nature of the overtones (p.50).

musical scale a series of musical notes arranged at regular intervals (↓), ranging from a low to a high pitch (↑).

wave of amplitude ● x
wave of amplitude ⦂ 2x
time
3rd harmonic
2nd harmonic
harmonic

harmonic (*n*) a musical note with a frequency that is equal to a whole number multiple of the fundamental frequency (p.50), e.g. the second harmonic has a frequency twice that of the fundamental frequency. When harmonics are added to a wave consisting of a single frequency, a **complex** wave is produced.

interval (*n*) (1) the distance between two objects; the time between two events. (2) the difference in the pitch of two notes; it is measured as the ratio of their frequencies, e.g. notes with pitches of 320 Hz and 256 Hz respectively have a frequency interval of 10:8.

octave (*n*) (1) the interval between any two frequencies that are in the ratio 2:1, e.g. musical notes with pitches of 480 Hz and 240 Hz are one octave apart. (2) the series of notes contained within this interval.

note	frequency Hz	pitch
C	512	high
B	480	
A	426.6	
G	384	
F	341.3	
E	320	
D	288	
C	256	low

octave (label spanning C through C)

octave

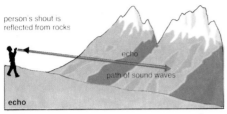

person's shout is reflected from rocks

echo

path of sound waves

echo

echo (*n*) a sound image produced when sound waves are reflected from a non-absorbing surface. **echo** (*v*).

echo sounder a device for measuring the depth of the sea from a ship. The echo sounder emits an ultrasonic (p.52) sound pulse which is reflected from the sea-bed. The sounder measures the time for the return of the echo and thus measures the depth of water.

echolocation (*n*) the location of an object by measuring the time for the return of an echo using sound waves or centimetre radio waves; used in sonar (↓) and radar (p.103).

sonar (*acro*) **so**und **na**vigation **r**anging. The detection of underwater objects by timing the return of an echo of an ultrasonic (p.52) sound pulse. An echo sounder (↑) is an example of sonar. Bats, when flying in the dark, use very high frequency sound waves to locate objects by echoes.

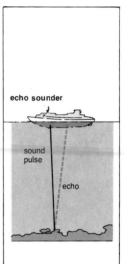

echo sounder

sound pulse

echo

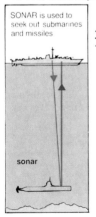

SONAR is used to seek out submarines and missiles

sonar

sonar

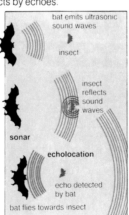

bat emits ultrasonic sound waves

insect

insect reflects sound waves

sonar

echolocation

echo detected by bat

bat flies towards insect

SOund **NA**vigation **R**anging

sonar

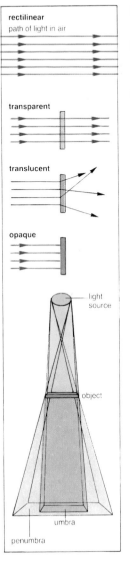

rectilinear
path of light in air

transparent

translucent

opaque

light
source

object

umbra

penumbra

source (*n*) the point of origin of energy, light, sound or materials, e.g. the Sun is a source of light and energy.

medium (*n. pl. media*) anything through which a force acts or an effect is transmitted. A medium may be matter, i.e. a solid, liquid or gaseous; or it may be a vacuum.

propagation (*n*) the transmission of a wave motion from a source (↑) by a medium (↑), e.g. the propagation of electromagnetic waves by a vacuum; the propagation of pressure waves by air. **propagate** (*v*).

rectilinear (*adj*) formed of straight lines, motion in a straight line, e.g. a square is a rectilinear shape; the path of light in air is rectilinear.

transparent (*adj*) describes any liquid or solid medium through which light can travel and form a sharp image. Clear glass and water are transparent media.

translucent (*adj*) describes a liquid or solid medium that is not transparent (↑) but allows light to pass through it; no clear image is formed, e.g. frosted glass is translucent but not transparent; some light passes through the glass, but the image formed is blurred.

opaque (*adj*) describes a solid or a liquid that does not allow light to pass through it, e.g. wood is opaque.

shadow (*n*) the dark area cast on a surface by an opaque (↑) object blocking rays of light. A shadow is formed because light travels in straight lines. The shadow will have the same shape as the object blocking the light rays, e.g. a square object will form a square shadow, a round object, a round shadow. The smaller the light source, the sharper will be the image.

umbra (*n*) the inner, darker area of the two areas seen in a shadow produced by a broad (extended) light source.

penumbra (*n*) the outer, lighter area of the two areas seen in a shadow produced by an extended light source. The penumbra is very narrow and is formed owing to some light reaching it from the source, because the source is broad. A very small point light source forms an umbra (↑) only, e.g. the light from a small torch.

fluorescence (*n*) a property of many substances of absorbing light or ultraviolet radiation and emitting light of a longer wavelength than that of the absorbed radiation. e.g. quinine sulphate absorbs u.v. light and fluoresces blue. The fluorescence stops immediately the radiation source is cut off. **fluoresce** (*v*). **fluorescent** (*adj*).

phosphorescence (*n*) a property of some substances of absorbing light or ultraviolet radiation and emitting light of a longer wavelength for a period of time after the radiation source is cut off. The length of time during which phosphorescence occurs varies with different substances. Calcium sulphide and zinc sulphide are examples of phosphorescent substances. **phosphoresce** (*v*). **phosphorescent** (*adj*).

primary colour the three colours, red, green and blue, are primary colours of light. Mixing all three by addition forms white light. Any other colour can be formed by a suitable mixture of all three colours by addition. Any two primary colours mixed by addition form a secondary colour (↓).

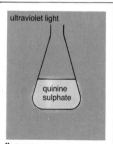

fluorescence

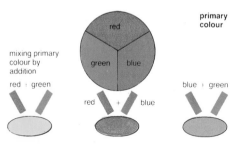

mixing primary colour by addition

primary colour

secondary colour the three colours, magenta, yellow and cyan, are secondary colours. Each is formed by mixing two primary colours (↑) by addition. Mixing two secondary colours by subtraction forms a primary colour.

complementary colour one of two colours, which, when mixed with the other colour by addition, forms white light, and when mixed by subtraction forms black, i.e. absorbs all light. For each primary colour there is a complementary secondary colour, and vice versa.

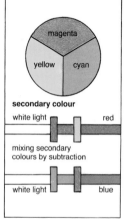

secondary colour

mixing secondary colours by subtraction

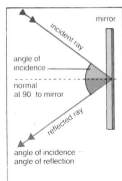

angle of incidence —

normal
at 90° to mirror

angle of incidence —
angle of reflection

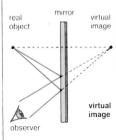

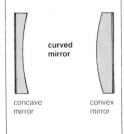

curved
mirror

concave
mirror

convex
mirror

reflect (*v*) to change the direction of a beam of radiation, such as light, sound or heat, by means of a surface, e.g. a mirror reflects light waves. **reflection** (*n*).

angle (*n*) a measurement of the change in direction between two straight lines that meet at a point, or two plane surfaces. Angles are measured in degrees or radians. There are 90 degrees (90°) or π in a right angle. **angular** (*adj*).

incident (*adj*) arriving at, falling upon or striking a surface. Incident rays of light are those rays hitting a surface; reflected rays of light are those rays leaving a surface.

mirror (*n*) a very smooth, polished surface capable of producing images by reflecting light from objects. A plane mirror produces an image that appears to come from a point as far behind the mirror as the object is in front; the image is also reversed left to right.

plane mirror see **mirror** (↑).

screen (*n*) (1) any flat surface on which an image can be formed by light rays. Examples of screens are a piece of white card or a sheet of white cloth. (2) a surface which acts as a shield, preventing light, sound or any unwanted effect from passing through it. **screen** (*v*).

ray (*n*) a straight line representing the direction of light, or other wave motion (p.49), from its source to any given point.

image (*n*) a copy of an object produced by a mirror or a lens (p.59), or formed on the retina (p.201) of the eye. An image may be larger, smaller or the same size as the object; it may be upright or inverted, real (↓) or virtual (↓).

real image an image formed at the point of the actual intersection of rays of light. Such images can be presented on a screen.

virtual image an image seen at a point from which rays of light appear to come. It cannot be presented on a screen as the light rays do not actually pass through the image, e.g. a mirror image is a virtual image.

curved mirror refers to a mirror with a surface which is either convex or concave. A concave mirror is capable of forming a real image from parallel rays of light, whilst a convex mirror forms a virtual image.

refraction (*n*) the change in direction of a ray (p.57) of light (or a wave of radiant heat or sound) as it passes from one medium (p.55) to another of a different density (p.10); the ray changes direction because its speed changes. **refract** (*v*), **refractive** (*adj*).

angle of incidence the angle between the entry point of an incident ray, and an imaginary line, the normal (↓), perpendicular (at a right angle) to the surface at that point.

angle of refraction the angle between the path of a refracted ray and an imaginary line, the normal (↓), perpendicular to the surface at that point.

normal (*n*) a line drawn perpendicular to the surface of a medium at a particular point, e.g. where a light ray strikes the surface.

Snell's law refers to the refraction of light rays; it states that for a ray of light refracted at the surface between two media, the ratio of the sine of the angle of incidence (↑) to the sine of the angle of refraction (↑) is a constant for those media.

refractive index an indication of the ability of a medium to refract light. It is the ratio of the sine of the angle of incidence of the ray passing from one medium to another, to the sine of the angle of refraction. In symbols, $n = \sin i / \sin r$. It is a constant for any two particular media, e.g. the refractive index of air to water is 1.33.

critical angle for a ray of light travelling from a dense medium to one of a lesser density, the critical angle is the smallest angle of incidence for which no refraction occurs. The angle of refraction would be 90°, but instead the ray is reflected by the surface of the medium.

total internal reflection occurs when a ray of light travelling through a dense medium, such as glass, is reflected back into that medium. Total internal reflection happens when the angle of incidence of the ray is greater than the critical angle (↑) of the medium.

mirage (*n*) the image produced by the reflection of light due to the circulation of layers of very warm air, heated by the Earth's surface. The image has a rippling appearance making it look like the surface of water; the movement is due to air currents.

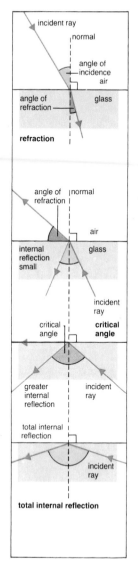

lens (*n. pl. lenses*) a piece of transparent material, such as plastic or glass, with one or both sides curved so that light can be refracted to form an image.

principal axis a line passing through the optical centre of a curved mirror or a lens and at a right angle to the mirror or lens.

focus (*n*) (1) a common point at which rays of light meet, and the position of the point can be located on a screen: this is a **real** focus. (2) a **virtual** focus is a point from which diverging rays from a lens appear to come. It cannot be located on a screen. (3) a **principal** focus of a lens is a point to which all parallel rays will be refracted or through which they will appear to pass.

focal length the distance between the principal focus of a lens and its centre.

converging lens a lens which refracts the rays of a beam of light and will bring parallel rays to a focus at a single point, i.e. the rays converge. A convex lens is a converging lens. It produces a real image at its principal focus.

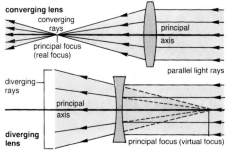

convex lenses

concave lenses

diverging lens a lens which refracts the rays of a beam of light and will cause parallel rays to spread out and appear to come from a point, i.e. the rays of light are diverged. A concave lens is a diverging lens, it forms a virtual image at its principal focus.

convex (*adj*) curved outwards. A lens may have only one convex surface, or both surfaces may be convex, e.g. a converging lens (↑).

concave (*adj*) curved inwards. A concave lens may have only one concave surface, or both surfaces may be concave, e.g. a diverging lens (↑).

camera (*n*) a box with a converging lens at one end and a light-sensitive **film** at the other, used to produce photographs by forming an image on the film. A **shutter** in front of the lens is opened to allow light to enter the camera to obtain a photograph. Objects at different distances are brought to focus on the film by moving the lens closer or further away from the film.

aperture (*n*) an opening. In a camera it is the hole through which light enters and forms an image on the film inside. The size of the aperture may be increased or reduced.

pinhole camera a very simple camera; it consists of a container with a film or screen at one end and a tiny hole at the other. Light from an object enters through the hole and forms an inverted real image (p.57) on the screen.

periscope (*n*) an optical instrument comprised of a large, thin tube fitted with lenses and reflecting mirrors or prisms; it is used for seeing objects that are above eye-level. Light from an object enters the periscope and is reflected down the tube by the mirror. At the bottom of the tube the light ray is reflected by another mirror and emerges travelling in its original direction, but at a lower level.

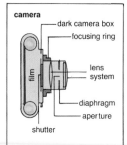

camera

dark camera box
focusing ring
lens system
diaphragm
aperture
film
shutter

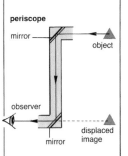

periscope

mirror
object
observer
mirror
displaced image

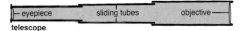

eyepiece — sliding tubes — objective
telescope

telescope (*n*) a device for viewing distant objects. It consists of two or more concentric tubes. Converging lenses at each end serve as the objective (↓) and the eyepiece (↓) to form a simple astronomical telescope.

binoculars (*n*) a device for viewing distant objects. It consists of two tubes (telescopes) one for each eye, both fitted with two converging lenses and with prisms to increase the distance between the lenses.

microscope (*n*) a device using two converging lenses in a special arrangement to magnify very small objects so that they can be seen. **microscopic** (*adj*).

eyepiece (*n*) the lens, or system of lenses, at the end of an optical instrument nearest the observer's eye.

objective (*n*) the lens, or system of lenses, at the end of an optical instrument nearest to the object that is being viewed.

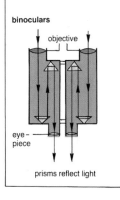

binoculars

objective
eye-piece
prisms reflect light

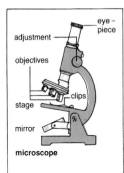

adjustment
eye – piece
objectives
clips
stage
mirror

microscope

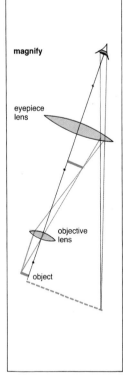

magnify

eyepiece lens

objective lens

object

magnify (*v*) to make an object appear larger when seen under a convex lens, or system of lenses, as in a microscope or telescope. **magnifying** (*adj*).

magnification (*n*) the ratio of the apparent size of the magnified object to its true size, i.e. of image to object.

reduce (*v*) to make an image smaller than the object.

prism (*n*) a regularly shaped block of transparent material used to reflect (p.57) or refract (p.58) light. Glass triangular prisms with two equal triangular faces joined by three rectangular, flat faces are most common.

deviation (*n*) a change in the direction of a ray of light or a wave motion when it passes from one medium to another. The angle through which the ray has been deviated can be measured.

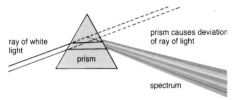

ray of white light

prism causes deviation of ray of light

prism

spectrum

spectrum (*n*) the result of splitting up light, by passing it through a prism, to give the colours: red, orange, yellow, green, blue, indigo, violet. The colours of the spectrum make up white light from the Sun.

rainbow (*n*) an arc in the sky formed from the colours of the spectrum. The effect is produced when bright sunlight shines on tiny water droplets. The sunlight is refracted, then internally reflected (p.57). Rainbows can be viewed only when the observer has his back to the Sun.

spectroscope (*n*) an instrument for producing and studying the colours of the spectrum. A simple spectroscope uses a glass prism or prisms to split white light into a spectrum, which can then be viewed through a telescope.

spectroscope

eyepiece

light enters through slit

draw tube

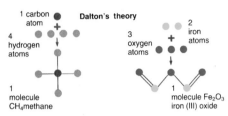

1 carbon atom

Dalton's theory

4 hydrogen atoms

1 molecule CH₄ methane

3 oxygen atoms

2 iron atoms

1 molecule Fe₂O₃ iron (III) oxide

Dalton's theory an explanation of the structure of elements and compounds. Every element is composed of atoms (p.45); the atoms of a particular element are identical, but are different from the atoms of every other element, if only in mass. Atoms of different elements join in simple numerical proportion to form compounds (p.45). Although modern theories disagree with Dalton's description of the atoms, his theory provides a useful explanation of the laws of chemical combination.

relative atomic mass the ratio of the average mass per atom of an element to $\frac{1}{12}$ of the mass of a carbon atom. The average mass is used because there may be isotopes (p.64) of the element.

relative molecular mass the ratio of the average mass per molecule of an element or compound to $\frac{1}{12}$ of the mass of a carbon atom. Relative molecular mass is equal to the sum of the relative atomic masses of all the atoms in the molecule.

combining weight a measure of the combining proportions of substances, relative to hydrogen as a standard. It is the mass, in grams, of an element that will combine with or replace 1 g of hydrogen or 8 g of oxygen.

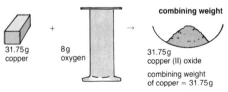

31.75 g copper

+

8 g oxygen

→

combining weight

31.75 g copper (II) oxide

combining weight of copper = 31.75 g

chemical equivalent (1) the combining weight. (2) of an acid (p.110), the mass of acid containing 1 g of replaceable acidic hydrogen; of a base (p.84), it is the mass required to neutralize (p.110) the equivalent mass of an acid.

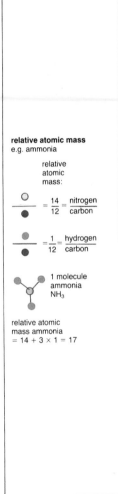

relative atomic mass
e.g. ammonia

relative atomic mass:

$\dfrac{\bigcirc}{\bullet} = \dfrac{14}{12} = \dfrac{\text{nitrogen}}{\text{carbon}}$

$\dfrac{\bullet}{\bullet} = \dfrac{1}{12} = \dfrac{\text{hydrogen}}{\text{carbon}}$

1 molecule ammonia NH₃

relative atomic mass ammonia
$= 14 + 3 \times 1 = 17$

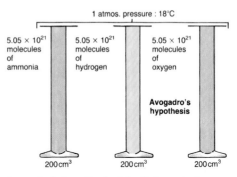

1 atmos. pressure : 18°C

5.05×10^{21} molecules of ammonia

5.05×10^{21} molecules of hydrogen

5.05×10^{21} molecules of oxygen

Avogadro's hypothesis

200 cm³   200 cm³   200 cm³

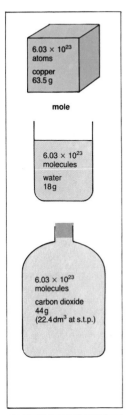

6.03×10^{23} atoms

copper
63.5 g

mole

6.03×10^{23} molecules

water
18 g

6.03×10^{23} molecules

carbon dioxide
44 g
(22.4 dm³ at s.t.p.)

Avogadro's hypothesis states that equal volumes of gases contain the same number of molecules, when measured under the same conditions of temperature and pressure.

mole (*n*) the SI unit for amount of substance. One mole is the amount of a substance that contains as many elementary units as there are atoms (p.45) in 0.012 kg of carbon–12, an isotope (p.64) of carbon. The elementary units must be specified, and can be atoms, molecules, electrons (p.64), ions (p.76), radicals (p.112) or other particles.

monatomic (*adj*) describes a molecule made up of one atom only, e.g. the inert (p.20) gases, such as helium and argon, are monatomic.

monatomic   ● diatomic   ●●

diatomic (*adj*) describes a molecule made up of two atoms, e.g. hydrogen gas H_2.

symbol (*n*) a sign, letter or diagram that represents something. In chemistry, letters are used as symbols of an element, or one atom of an element, e.g. O = oxygen; Na = sodium; Mg = magnesium.

formula (*n*) chemical symbols, written together, which show the combination of the atoms of each element in a molecule of a compound or ion, e.g. the formula H_2O stands for a molecule of water, and shows that each molecule comprises two hydrogen atoms combined with one atom of oxygen.

symbol	meaning
Na	sodium ●
Zn	zinc ●
Cl	chlorine ●
V	volume
p	pressure
T	temperature
≡	equivalent to
×	multiply

(● 1 atom/1 mole)

particle (*n*) any small piece of material, such as a grain of sugar, a molecule (p.24) or atom (p.45).

electron (*n*) a very small particle possessing a negative electric charge and forming a part of all atoms. Electrons are located in shells (↓) which orbit the nucleus (↓) of an atom; they can also be released from atoms and travel by themselves. An electron has a tiny mass, approximately 1/1840 that of a hydrogen atom.

proton (*n*) a tiny particle in the nucleus (↓) of an atom. It has a positive electric charge equal to the negative charge of an electron. Its mass is approximately 1840 times that of an electron. A proton is a hydrogen ion, (i.e. a normal hydrogen atomic nucleus) and is present in all atoms.

neutron (*n*) a tiny particle in the nucleus of all atoms, except hydrogen. It has no electric charge, and a mass only very slightly larger than that of a proton.

nucleus[1] (*n. pl. nuclei*) the central core of an atom, comprised of neutrons (↑) and protons (except for hydrogen which has a nucleus of only one proton). The nucleus has a positive electric charge. In the case of atoms, this charge is exactly the same magnitude as the negative charge of the orbiting electrons. Almost the whole of the mass of an atom is concentrated in its nucleus, as the electrons have such a tiny mass that it is negligible.

isotopes (*n*) atoms of the same element which contain the same number of protons in their nucleus, i.e. the same atomic number, but have a different number of neutrons, i.e. a different atomic mass. The isotopes of an element all possess the same chemical properties. Almost all elements occuring in nature are mixtures of several isotopes.

electron shell the sphere in which an electron is located in its orbiting path around the nucleus of an atom. There are a number of concentric shells around a nucleus, each a specific distance from the nucleus, and each capable of containing electrons up to a specific number.

K-shell (*n*) the shell of electrons closest to the nucleus; it can contain up to 2 electrons.

L-shell (*n*) the shell of electrons second closest to the nucleus; it can contain up to 8 electrons.

M-shell (*n*) the shell of electrons next to the L-shell; it can contain up to 18 electrons.

carbon-12 atom

electron

nucleus containing 6 protons + 6 neutrons

nucleus

K-shell

L-shell

M-shell

electron shell

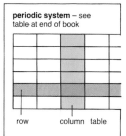

periodic system – see table at end of book

row   column   table

atomic number the number of protons in the nucleus of an atom of an element. The symbol is Z.

periodic system a table of the elements arranged in **rows** of increasing atomic number, and ordered in such a way that elements possessing similar physical and chemical properties form groups in **columns**.

transition elements elements of similar chemical properties which form the middle sections of rows in the periodic system. The elements are metals with more than one electrovalency ($\downarrow$) in which an inner shell of electrons also provides valency electrons.

bond (n) the force which holds atoms together to form molecules. The bond is formed when two atoms share two electrons (covalent bond) or an electron is transferred from one atom to another (electrovalent bond).

valency electron an electron in the outermost shell of an atom; it is used in forming chemical bonds.

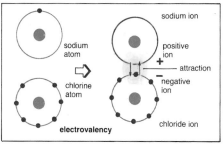

electrovalency (n) (1) the union of atoms by the transfer of electrons from one atom to another. The atom that loses one or more electrons becomes positively charged. The atom that gains electrons becomes negatively charged. The unlike charges attract each other and the two atoms are held together by an electrovalent bond. (2) a measure of the number of electrons that an atom has available to form electrovalent bonds. **electrovalent** (*adj*).

covalency (n) (1) the union of atoms by the sharing of one or more pairs of electrons, each atom containing one of the electrons of the pair. The pairs of electrons form a covalent bond. (2) a measure of the number of covalent bonds an atom can form.

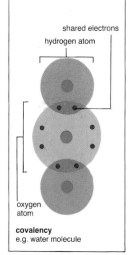

shared electrons
hydrogen atom

oxygen atom

covalency
e.g. water molecule

radioactivity (*n*) the property of spontaneous
disintegration of the nucleus (p.64) of an atom (p.45)
with the emission of alpha or beta particles (↓) or
gamma rays (↓). Radioactivity may occur naturally,
or it may be induced, it is not affected by external
conditions.

disintegrate (*v*) (1) to break up into fragments.
(2) the emission of a particle, or particles, from the
nucleus of a radioactive atom. **disintegration** (*n*).

alpha particle a combination of two neutrons and
two protons; it is identical to the nucleus of a helium
atom. Alpha particles have a positive charge, and
are emitted from certain radioactive substances.

beta particle a high velocity electron, with a range in
air of approximately 750 cm, emitted from a
radioactive nucleus. Disintegration of a neutron in
the nucleus produces a proton, and an electron
which is emitted as a beta particle.

range (*n*) (1) the furthest distance that anything can
travel or be ejected, e.g. the range of an alpha
particle in air is about 7 cm. (2) the values between
the upper and lower limits of a scale of
measurements.

gamma rays a highly penetrating form of radiation
which is emitted with alpha or beta particles from
the nucleus of certain radioactive atoms. The rays
are similar to X-rays, but have a shorter wavelength.
They travel at the speed of light, and have no
electric charge.

half-life the time taken for the disintegration of half of
the atoms in a given sample of a radioactive
substance. Each radioactive element has a definite
half-life, and it is an important characteristic which
helps to identify the element.

nuclear fission the disintegration of a heavy atomic
nucleus into two smaller, approximately equal
nuclei; at the same time neutrons are emitted and
huge quantities of energy are released. Fission may
be spontaneous, or it may be caused when high
energy particles (neutrons) bombard a nucleus or
nuclei.

fission (*n*) nuclear fission (↑).

nuclear fusion the union of two light atomic nuclei to
form one heavy nucleus; there is an overall loss of
mass which results in the release of a large quantity
of energy.

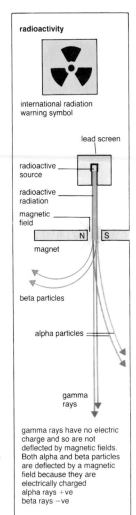

radioactivity

international radiation
warning symbol

lead screen

radioactive
source

radioactive
radiation

magnetic
field

N   S

magnet

beta particles

alpha particles

gamma
rays

gamma rays have no electric
charge and so are not
deflected by magnetic fields.
Both alpha and beta particles
are deflected by a magnetic
field because they are
electrically charged
alpha rays +ve
beta rays −ve

nuclear energy the energy released in nuclear fission (↑). When 1 g of uranium–235 undergoes nuclear fission, 8.2×10^7 kJ of energy are released.

atomic energy common term for nuclear energy (↑). Nuclear energy is more correct.

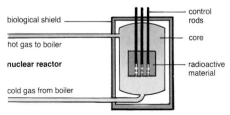

control rods

biological shield

core

hot gas to boiler

nuclear reactor

radioactive material

cold gas from boiler

nuclear reactor an industrial plant for the controlled release of nuclear fission energy. It has an active core containing radioactive material with the fission controlled by rods of a substance which readily absorbs neutrons, e.g. boron-steel or cadmium metal. The nuclear fission develops heat which is removed by a cooling system, such as carbon dioxide, helium or molten sodium, flowing round pipes. The heat is used to produce steam which drives a turbine. The control rods are moved down into the core to reduce nuclear fission, or raised to increase it. The reactor is surrounded by a biological shield of thick concrete which absorbs any stray radiation.

reactor (*n*) a nuclear reactor (↑).

Geiger-Müller tube an instrument which is used for detecting and measuring radiation, chiefly alpha, beta or gamma rays. Rays entering the counter cause discharge between the anode and cathode electrodes. The discharge is used to produce a noise and trigger an automatic counting device.

fallout (*n*) particles of radioactive substances which fall onto the Earth from the atmosphere after the explosion of a nuclear device, e.g. an atomic bomb. **Local** fallout consists of large particles falling within a radius of 150 km during the first few hours. 'Tropospheric fallout' consists of fine particles falling approximately along the same latitude during the first week or so. **Stratospheric** fallout consists of all types of particles falling over the world's surface over a period of years.

Geiger-Müller tube

thin mica window

holder

anode

cylindrical metal cathode containing inert gas

coaxial lead for connection to indicating equipment

magnet (*n*) a solid object that possesses the property of magnetism; it attracts iron, and attracts or repels other magnets. It will lie in a north-south direction when it is free to turn. **magnetize** (*v*), **magnetic** (*adj*).

permanent magnet a magnet that retains its magnetism after it has been magnetized. It is usually an alloy of steel.

temporary magnet a material, usually of soft iron, that displays the properties of a magnet when it is in a magnetic field, but loses its magnetism when the field is removed.

pole (*n*) one of the ends of a magnet at which the magnetism appears to be concentrated. The north-seeking pole is the end of the magnet which points north when the magnet is free to turn; the south-seeking pole is the end which points south.

keeper (*n*) a soft iron bar placed against the poles of a permanent horseshoe magnet, or the opposite poles of two adjacent bar magnets, to prevent loss of magnetism.

magnetic field the space in which a magnet or an electric current is capable of exerting a force of attraction on magnetic materials, or the force which causes a magnet to set itself in a particular direction.

terrestrial magnetism refers to the Earth's magnetic field; it is similar to a field that would be produced by a powerful magnet situated in the Earth's core.

compass (*n*) an instrument that indicates direction. It is comprised of a freely pivoted magnet contained in a case. Below the magnet is a scale marked with the directions north, east, south and west.

magnetic dip the angle between the direction of the Earth's magnetic field and the horizontal at any point on the Earth's surface.

magnetic inclination *see* **magnetic dip**.

dip circle an instrument which measures the angle of magnetic dip (↑). It consists of a magnetized needle free to turn in a vertical plane, the angle of dip being measured on a circular scale.

magnetic declination the angle between magnetic north, as shown by a compass needle, and true north. Magnetic declination varies according to the position of the compass on the Earth's surface.

magnetic variation *see* **magnetic declination**.

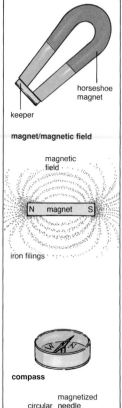

horseshoe magnet

keeper

magnet/magnetic field

magnetic field

N magnet S

iron filings

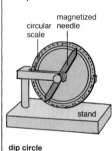

compass

magnetized needle

circular scale

stand

dip circle

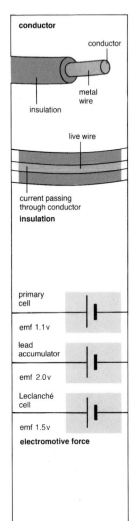

conductor

conductor

metal
wire

insulation

live wire

current passing
through conductor
insulation

primary
cell

emf 1.1 v

lead
accumulator

emf 2.0 v

Leclanché
cell

emf 1.5 v

electromotive force

current (*n*) (1) the flow or rate of flow of electric
charge through a conductor (↓). (2) the movement
of a fluid in a particular direction, e.g. air currents,
ocean currents.

conduct (*v*) to allow electric current or heat to flow
through a material. **conduction** (*n*) the action of
conducting.

conductor (*n*) a solid, liquid or gas that readily
allows an electric charge to flow through it, or
readily conducts heat, sound etc. Metals are very
good conductors of heat and electricity.

insulator (*n*) a substance or an object that does not
readily allow the flow of an electric current or heat.
Most non-metals, such as plastics, and all gases
are insulators.

insulate (*v*) to cover with a material that prevents
electricity or heat entering or leaving.

insulation (*n*) (1) the process of insulating. (2) any
material through which an electric current or heat
cannot pass, e.g. rubber acts as insulation for
electric cables.

electric (*adj*) describes a device that produces,
uses or conveys an electric current, e.g. an electric
motor, an electric circuit.

electrical (*adj*) to do with the subject of electricity,
e.g. electrical engineer; or concerned with the
operation of electricity, e.g. electrical applicance.

live (*adj*) describes a wire or cable (p.73) through
which an electric current is flowing. It usually
implies that it is dangerous to touch the wire or
cable, as the current is sufficiently strong to cause
shock.

electromotive force (e.m.f.) the force, measured in
volts, that pushes an electric current around an
electrical circuit. A battery or a generator (p.208)
produces an electromotive force; the e.m.f. is
constant for any one kind of cell so long as
polarization (p.70) does not occur. The symbol for
electromotive force is *E*.

potential difference (pd) the difference in the
electromotive force between any two points, which
results in the flow of electric current between them.
Current always flows from points of higher potential
to those of a lower potential, conventionally this is
taken to be from a positive to a negative potential.
Potential difference is measured in volts V.

resistance² (*n*) a property of substances that reduces, or acts against the flow of an electric current through a conductor or insulator. The resistance of an insulator is very high. A thin metal wire will offer more resistance to an electric current than a thick wire of the same metal. Resistance is measured in ohms. Symbol for resistance is *R*.

primary cell a device capable of producing an electric current as a result of a chemical reaction. A simple cell consists of two different metals joined by a wire conductor, and immersed in an acid, or alkali or any solution of an electrolyte.

polarization (*n*) the formation of bubbles of hydrogen around the positive metal plate of a primary cell (↑). Gas is a good insulator (p.69), and therefore the bubbles of air reduce the electric current flowing through the cell. **polarize** (*v*).

depolarization (*n*) preserving the efficient operation of a primary cell by preventing polarization. Depolarization entails adding a substance which chemically combines with the hydrogen in the acid of the cell, thereby preventing the formation of hydrogen bubbles. The substance used is called a **depolarizer**. **depolarize** (*v*).

local action the process of bubbles of hydrogen forming on the zinc plate of a primary cell, thereby reducing its efficiency. Local action is caused by impurities in the zinc and is prevented by rubbing the zinc with mercury.

Leclanché cell a type of primary cell (↑). It consists of a positive carbon rod packed in a porous pot with a depolarizer (manganese dioxide), and standing with a negative zinc rod in a solution of ammonium chloride.

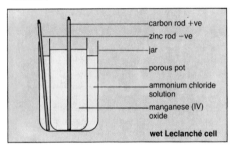

wet Leclanché cell

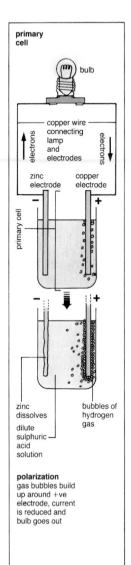

primary cell

polarization
gas bubbles build up around +ve electrode, current is reduced and bulb goes out

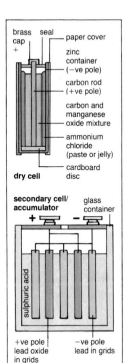

brass seal
cap
+
paper cover
zinc
container
(−ve pole)
carbon rod
(+ve pole)
carbon and
manganese
oxide mixture
ammonium
chloride
(paste or jelly)
cardboard
disc

dry cell

**secondary cell/
accumulator**
glass
container

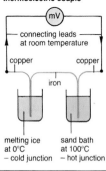

sulphuric acid

+ve pole
lead oxide
in grids

−ve pole
lead in grids

thermoelectric couple

mV

connecting leads
at room temperature

copper   copper

iron

melting ice
at 0°C
− cold junction

sand bath
at 100°C
− hot junction

dry cell a Leclanché cell which uses a paste of
ammonium chloride instead of a solution. Dry cells
can only be used for short periods at a time.

battery (*n*) two or more primary or secondary cells
used together to produce or store electricity.

dry battery a battery made up of one or more dry
cells (↑).

secondary cell an electric cell in which a chemical
reaction is used to produce an electric current. A
secondary cell is recharged by passing an electric
current in the opposite direction to discharge. A
secondary cell produces and stores electricity,
whereas a primary cell (↑) only produces electricity.

storage cell an alternative name for a **secondary
cell**.

accumulator (*n*) a group of secondary cells used for
storing electrical energy; the energy is discharged
when needed. Accumulators are used in car
engines to start the motor.

Seebeck effect if two wires of different metals are
joined to form a circuit, and the two junctions are
maintained at different temperatures, an electric
current flows around the circuit. This is called the
Seebeck effect.

iron

Seebeck effect

hot
junction

cold
copper   junction

thermoelectric (*adj*) to do with the production of
electrical energy directly from heat energy.

thermoelectric couple a pair of two different metal
wires joined at two junctions, and used in the
Seebeck effect (↑). The junction of the wires is
called a **thermoelectric junction**.

thermocouple (*n*) an alternative name for
thermoelectric couple.

thermopile (*n*) an instrument for measuring heat
radiation. It is comprised of a series of
thermocouples (↑); when their junctions are
exposed to heat an electric current is produced
which is then measured by a galvanometer (p.74).

Peltier effect the warming or cooling of the junction
of two wires of different metals is dependent on the
direction of flow of the electric current, and the total
charge crossing the junction.

ampere (*n*) the SI unit of electric current. A current of 1 ampere flowing through each of two parallel conductors (p.69) placed 1 metre apart in a vacuum will produce a force of 2×10^{-7} newtons per metre length of the conductors. The symbol for ampere is A.

coulomb (*n*) the SI unit of electric charge, measured as the quantity of charge transferred by 1 ampere in 1 second. The symbol is C. A current of 12 A flowing for 2 seconds transfers 24 C.

volt (*n*) the SI unit of electric potential. It is measured as the potential difference (p.69) between two points in an electric circuit that requires 1 joule of work to move a charge of 1 coulomb from the point of lower potential to that of higher potential. The symbol is V.

ohm (*n*) the SI unit of resistance, defined as the resistance between two points in a circuit when a potential difference (p.69) of 1 volt between the points maintains a current of 1 ampere. The symbol is Ω.

circuit (*n*) a complete path over which an electric current can flow. It is generally comprised of a series of conductors and components offering resistance to the current. It is also known as a **closed circuit**; current can flow from the source of electrical energy around the circuit and back to the source.

open circuit a circuit which has been broken at one point, and around which a current cannot flow.

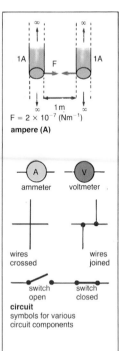

$$F = 2 \times 10^{-7} \ (Nm^{-1})$$
ampere (A)

ammeter   voltmeter

wires
crossed

wires
joined

switch
open

switch
closed

circuit
symbols for various
circuit components

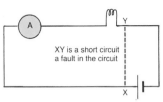

XY is a short circuit
a fault in the circuit

X

short circuit

short
circuit

short circuiting an
ammeter

short circuit (1) an accidental fault in an electrical circuit which causes a large current to flow through a battery, or other electrical source. A connection is made which provides the current with a path of low resistance, as other parts of the circuit are cut off from the supply of current. (2) an intentional connection to remove a component (↓) from a circuit, e.g. to short circuit a galvanometer.

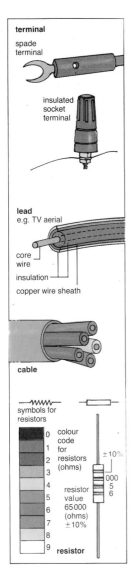

terminal

spade
terminal

insulated
socket
terminal

lead
e.g. TV aerial

core
wire

insulation

copper wire sheath

cable

symbols for
resistors

colour
code
for
resistors
(ohms)

0
1
2
3
4
5
6
7
8
9

±10%

000
5
6

resistor
value
65000
(ohms)
±10%

resistor

component[2] (n) an electrical device in a circuit, e.g. a resistor, lamp, switch, capacitor, etc. A cell or battery is not considered to be a component.

terminal (n) a device to which a wire can be attached to join different parts of an electric circuit. The wire is connected to the terminal by means of a nut on a screw, or by a clip.

lead (n) a wire making an electrical connection between two components (↑) of a circuit or between two parts of a circuit. It is usually covered with an insulating material, e.g. the lead from the aerial to a television set.

flex[1] (n) a flexible electric wire that is insulated. Flex is used to connect parts of an electric circuit.

cable (n) a thick wire, made from several wires twisted together, covered with an insulating material, such as rubber, and sometimes an outside cover of cloth or plastic. It is used for carrying large electric currents, e.g. a cable supplying current to an electric motor; a cable carrying current to a building. Overhead cables are used for electricity supplies.

mains (n.pl.) the cables, transformers and distribution points supplying domestic electricity from power stations to houses and factories.

switch (n) a device used to open and close an electric circuit, e.g. when an electric lamp switch is on the circuit is closed and current flows, when the switch is off the circuit is open and the current cannot flow.

resistor (n) a device used in an electric circuit to provide resistance to the current. A **fixed resistor** has a specific value expressed in ohms.

variable resistor a resistor whose resistance can be altered to cover a range of values; this is done by altering the length of the material through which the current must travel.

rheostat (n) see **variable resistor** (↑).

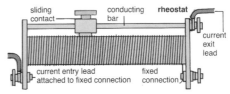

sliding
contact

conducting
bar

rheostat

current
exit
lead

current entry lead
attached to fixed connection

fixed
connection

capacitor (*n*) a device which stores electric charge. It consists of two conductors separated by insulating material. The simplest capacitor is two parallel metal plates separated by air (an insulator). Cheap capacitors consist of two sheets of metal foil separated by a sheet of paper, the whole rolled into a cylinder, and put in a plastic cover.

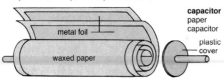

capacitor
paper
capacitor

metal foil

waxed paper

plastic
cover

heating element a wire of low resistance used to produce heat in an electrical appliance, such as an electric iron or electric kettle. When an electric current flows through it, the wire produces heat.

filament[1] (*n*) a very fine wire of high resistance which glows white-hot when an electric current is passed through it. Filaments are used in electric light bulbs, and in thermionic valves (p.87).

fuse (*n*) a special safety device consisting of a thin wire which melts and breaks the circuit if too great an electric current passes through it.

series circuit a circuit in which the components are so connected that the current flows through each device in turn.

parallel circuit a circuit that is so arranged that the current divides into two or more paths between two points in the circuit. The potential difference across each path is the same.

Ohm's law at a constant temperature, the intensity of the current flowing through a metal conductor is directly proportional to the potential difference between the ends of the conductor. The resistance of the conductor is measured as the potential difference divided by the strength of the current.

galvanometer (*n*) an instrument for detecting and measuring small electric currents. It does not measure current in amperes, only in scale units.

ammeter (*n*) a low resistance instrument that measures the electric current strength, in amperes.

voltmeter (*n*) a high resistance instrument for measuring the potential difference (p.69) in volts between any two points in a circuit.

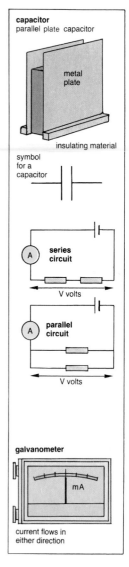

capacitor
parallel plate capacitor

metal
plate

insulating material

symbol
for a
capacitor

**series
circuit**

V volts

**parallel
circuit**

V volts

galvanometer

mA

current flows in
either direction

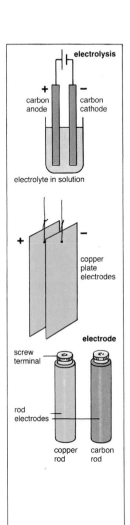

electrolysis

carbon anode +
carbon cathode −

electrolyte in solution

+ −

copper plate electrodes

electrode

screw terminal

rod electrodes

copper rod   carbon rod

electrolysis (*n*) the chemical decomposition of a dissolved or molten substance, caused by passing an electric current through the liquid.

voltameter (*n*) a device for measuring the quantity of current that has passed through a circuit. It is calculated by measuring the quantity of metal deposited, or the amount of gas liberated during a definite period of electrolysis.

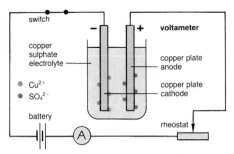

switch

voltameter

copper sulphate electrolyte

copper plate anode

Cu²⁺
SO₄²⁻

copper plate cathode

battery

rheostat

electrolytic cell a container in which electrolysis occurs.

electrode[1] (*n*) a piece of conducting material through which a current enters or leaves an electrolytic cell during electrolysis (↑).

anode (*n*) the positively charged electrode (↑) towards which negative ions (p.76) flow during electrolysis.

cathode (*n*) the negatively charged electrode towards which positive ions (p.76) move during electrolysis.

electrolyte (*n*) a compound that, when molten or in solution, will conduct an electric current and is chemically decomposed during the process. The current is carried by ions (p.76).

non-electrolyte (*n*) a substance that does not produce ions (p.76) in solution, and therefore will not conduct electricity. Ethanol is a non-electrolyte.

Faraday's laws of electrolysis (1) during electrolysis, the amount of chemical action, i.e. the mass of substances deposited or liberated, is proportional to the strength of the current and its duration. (2) the masses of substances deposited or liberated by a specific quantity of electric charge (measured in coulomb) are proportional to their chemical equivalents (p.62), e.g. 96 500 coulomb of current will liberate 1 g of hydrogen, or 108 g of silver.

ion (*n*) a positively or negatively charged atom (p.45) or group of chemically combined atoms which carry electric charge during electrolysis (p.75). *See* **anion** (↓) and **cation** (↓). **ionic** (*adj*).
ionize (*v*) to change into ions. **ionization** (*n*).
anion (*n*) a negatively charged ion, formed when an atom or group of atoms gains one or more electrons. Anions move towards the anode (p.75) during electrolysis, e.g. bromine ion Br⁻

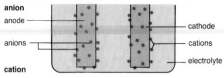

anion
anode
anions
cation
cathode
cations
electrolyte

cation (*n*) a positively charged ion formed by the loss of one or more electrons from an atom or group of atoms. Cations move towards the cathode (p.75) during electrolysis, e.g. silver ion Ag⁺
deposit (*n*) the metal layer laid down on the cathode during electrolysis, e.g. a layer of silver deposited on the cathode in the electrolysis of silver nitrate.
faraday (*n*) a unit of quantity of electric charge; 1 faraday is the quantity of charge needed to free 1 g of hydrogen or 108 g of silver; it is the charge on 1 mole of electrons. It is equal to about 96 500 coulomb. Symbol: *F*.
electrochemical equivalent the mass of a substance liberated or deposited at an electrode during electrolysis by the passage of 1 coulomb of electric charge.
electroplating (*n*) the deposit of a coating of metal on another metal by electrolysis. The metal to be coated forms the cathode, and the salt of the metal that is to be deposited is contained in solution.

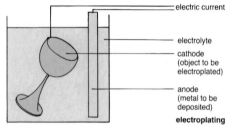

electric current
electrolyte
cathode (object to be electroplated)
anode (metal to be deposited)
electroplating

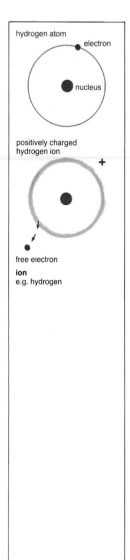

hydrogen atom
electron
nucleus

positively charged hydrogen ion
+

free electron
ion
e.g. hydrogen

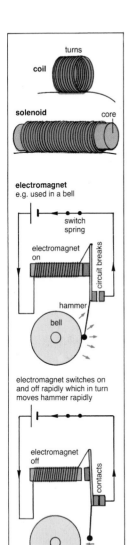

turns

coil

solenoid

core

electromagnet
e.g. used in a bell

switch
spring

electromagnet
on

circuit breaks

hammer

bell

electromagnet switches on
and off rapidly which in turn
moves hammer rapidly

electromagnet
off

contacts

electromagnetism (*n*) (1) the magnetism produced by an electric current flowing through a wire wound around an iron bar. (2) the subject concerned with electricity and magnetism.

coil (*n*) wire wound in rings around a rod (a former), or just wound in a spiral. When an electric current is passed through a coil of wire, a magnetic field is produced.

solenoid (*n*) a coil of wire around a soft iron core. The solenoid acts like a magnet when the wire carries an electric current.

turns (*n*) the number of circles of wire wound around a rod (a former).

winding (*n*) (1) the action of coiling wire around a rod. (2) a turn, or turns, of wire wound around a rod, e.g. the winding on an electric coil.

core (*n*) the central part of an object. In electromagnetism, the core is the iron rod at the centre of a coil or solenoid that helps to channel the magnetic field, or it is the space in the middle of the coil when there is no bar or rod.

electromagnet (*n*) a temporary magnet made by winding a coil around a soft iron core. It only acts as a magnet when there is a current flowing through the coil.

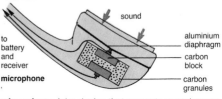

sound

to
battery
and
receiver

microphone

aluminium
diaphragm

carbon
block

carbon
granules

microphone (*n*) a device that converts sound waves into electrical energy. A **carbon microphone** consists of a diaphragm that is moved backwards and forwards by sound waves. The movement varies the pressure on carbon granules between a front, movable carbon block and a back, fixed carbon block. When the granules are compressed (because the pressure on the diaphragm increases) their electrical resistance decreases. The opposite occurs when there is a decrease in pressure. The variation in the resistance of the granules causes a variation in the current passing through the microphone to a receiver (p.78).

receiver (*n*) a device that converts incoming electrical signals into sound waves, or pictures. A diaphragm is positioned above a permanent magnet around which coils of wire are wound. The diaphragm vibrates due to any varying electric current (passed from the microphone) and produces corresponding sound waves.

loudspeaker (*n*) a type of receiver (↑) producing sounds loud enough to be heard at a distance. Most loudspeakers use a moving-coil mechanism. A coil, wound on stiff paper, moves in and out of a circular magnet. The electrical signals cause the coil to vibrate a paper cone and reproduce sound.

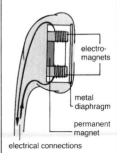

receiver
(in a telephone)

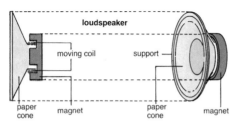

amplify (*v*) to increase the magnitude of a wave form of energy, e.g. sound waves. To increase the magnitude of an electric current or voltage, e.g. a loudspeaker (↑) circuit amplifies the original electric current taken in by a microphone; a transistor amplifies an electric current. **amplification** (*n*).

relay (*n*) a device which uses an electric current in one circuit to turn a switch on or off, and thereby control the electric current in another circuit.

telephone (*n*) a device for sending sounds over long distances. It consists of a microphone that changes speech into electrical impulses. The impulses are carried by wires to a receiver which changes the electrical impulses back into speech.

telegraph (*n*) a device for sending messages over a distance by sending electrical impulses along a wire. When a key is pressed at the transmitting end it closes the circuit and a current flows to the receiver. The current only flows when the key is pressed. The intermittent current operates a relay whose second circuit operates a buzzer, or records the dots and dashes of the Morse code.

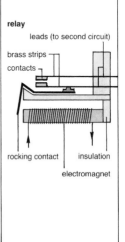

Morse code

dot ●
dash ━

letter	R	O	S	E
Morse code	● ━ ●	━ ━ ━	● ● ●	●

Morse code a code using long and short signs,
called dashes and dots. The code was first used on
telegraphs. Each letter and number has its own
combination of dots and dashes. Morse code can
be used with any form of signalling equipment
which can make long and short units, e.g. flashing a
lamp; waving a flag; a knock on a hard material.

moving-coil (*adj*) describes any instrument that
uses the rotation of a coil in a magnetic field. In a
moving-coil ammeter the coil is pivoted between
the poles of a permanent magnet. When current
flows through the coil it sets up a magnetic field,
and the coil moves in reaction to the poles of the
permanent magnet. A pointer attached to the coil
indicates the magnitude of the electric current on a
scale graduated in amperes.

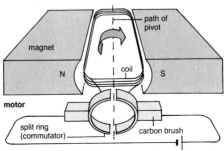

path of
pivot

magnet

N coil S

motor

split ring
(commutator)

carbon brush

motor[1] (*n*) a machine, other than an engine, that
changes one form of energy into mechanical
energy. An electric motor transforms electrical
energy into mechanical energy. In its simplest form
it consists of an armature (↓) through which an
electric current flows, placed between the two poles
of a powerful magnet. The armature rotates
producing mechanical energy.

armature (*n*) a piece of soft iron with a coil wound
around it, which is free to rotate on an axle. When an
electric current is passed through the coil the
armature of a motor rotates producing mechanical
energy.

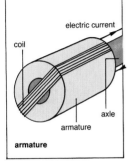

electric current

coil

axle

armature

armature

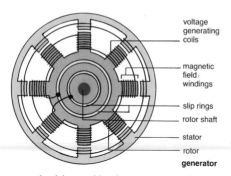

voltage
generating
coils

magnetic
field
windings

slip rings

rotor shaft

stator

rotor

generator

generator (*n*) a machine that converts mechanical energy into electrical energy. It relies on the principle that if an electric conductor is moved in a certain direction in a magnetic field, an electric current is induced. A simple generator consists of an armature placed between the poles of a permanent magnet. When the armature is turned (mechanical energy), a current (electrical energy) flows in the coil.

dynamo (*n*) an alternative name for a **generator** (↑).

alternator (*n*) a generator (↑) that produces alternating current (↓).

rotor (*n*) the rotating part of a motor, generator or turbine (p.19).

stator (*n*) the stationary part of a motor, generator or turbine (p.19).

hydroelectric (*adj*) describes the generation of electricity by the use of water power, e.g. a flow of water drives a turbine which drives a generator.

direct current an electric current that always flows in the same direction.

alternating current an electric current that changes its direction many times per second. The current travels in one direction, increasing until it reaches a maximum, it then decreases and changes direction. The pattern is then repeated in the other direction. The frequency is the number of complete cycles per second. A complete cycle is the movement of current to the furthest point in one direction and back to the furthest point in the opposite direction.

rectifier (*n*) a device that transforms an alternating current into a direct current.

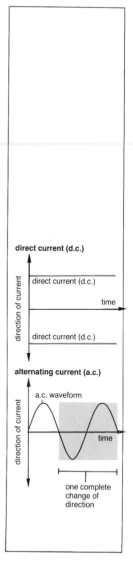

direct current (d.c.)

direction of current

direct current (d.c.)

time

direct current (d.c.)

alternating current (a.c.)

direction of current

a.c. waveform

time

one complete
change of
direction

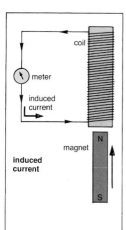

induced
current

smooth (v) a rectifier (↑) produces a voltage with a series of peaks. To overcome this, the voltage is smoothed to produce only a small variation in voltage, called a **ripple**.

filter[1] (n) a device using a capacitor (p.74) and a resistor to remove certain frequencies of an alternating current. In a high-pass filter, frequencies below a certain value are reduced or removed. A low-pass filter reduces or removes frequencies above a certain value. A simple filter circuit does not have a sharp cut-off point for frequencies.

induced current refers to a current which flows in a conductor when it is moved in a magnetic field, or when the strength of the magnetic field is altered.

Faraday's law an electromotive force (p.69) is induced whenever the magnetic field passing through a conductor is altered; the strength of the e.m.f. is proportional to the rate of change of the strength of the magnetic field.

Lenz's law when a current is induced in a circuit it will flow in the direction that produces a magnetic field opposing the motion that induced it.

transformer (n) a device used to change the voltage of an alternating current (↑) without changing its frequency. It is comprised of a soft iron core, on which is wound a primary coil consisting of a few turns of thick wire. Over the primary coil is wound a secondary coil comprised of many turns of very thin wire. An alternating current is passed through the primary coil, and an induced alternating current is obtained from the secondary coil.

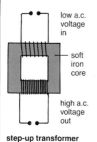

step-up transformer

turns ratio in a transformer it is the number of turns on the secondary coil divided by the number of turns on the primary coil, e.g. 2000 secondary turns and 50 primary turns gives a turns ratio of 2000/50 = 40, i.e. 40:1.

step-up transformer a transformer (↑) in which the voltage produced in the secondary coil is greater than that in the primary coil. It has a turns ratio (↑) greater than 1, i.e. it has more turns in the secondary than in the primary coil.

step-down transformer a transformer (↑) in which the voltage produced in the secondary coil is less than that in the primary coil. It has a turns ratio of less than 1, i.e. it has fewer turns in the secondary than in the primary coil.

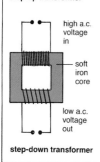

step-down transformer

electrostatic (*adj*) describes electric charges that are stationary (they do not flow), and their effects.

electric charge a property possessed by certain tiny particles of matter. The nature of an electric charge is unknown, but its effects can be observed: a charge will be repelled by a like charge, and attracted by an unlike charge; the field in which these forces of attraction and repulsion operate is known as an electric field; when a charge is discharged (escapes to earth) a spark is produced. There are two types of charge; positive and negative. **charge** (*v*), **charged** (*adj*), **discharge** (*n*), **discharge** (*v*).

positive (*adj*) describes one type of electric charge. Originally the charge produced on glass by rubbing it with silk was called positive, and the charge on the silk was called negative. The choice of positive and negative for the two types of charge was solely a convention; as there are two opposite types of charge, once positive is decided, the other type has to be negative. A positive potential is considered higher than a negative potential, hence conventional current flows from positive to negative. In reality, electrons (p.64) flow from negative to positive.

negative (*adj*) *see* **positive** (↑).

spark (*n*) a short, rapid discharge of an electric current through a gas, such as air. It is accompanied by sound and a flash of light.

lightning (*n*) a very powerful electric spark. It is caused by the discharge between an electrostatically charged cloud and the ground, or between two charged clouds.

thunder (*n*) the sound produced by a charged cloud suddenly heating and expanding the air when it is discharged.

lightning conductor a cloud moving through air becomes charged owing to friction between air and the cloud. A lightning conductor has a set of sharp spikes to aid discharge. The cloud induces an opposite charge on the spikes. The spikes readily lose their charge and this discharges the cloud as the opposite charges neutralize each other. As the cloud is discharged, there is no lightning.

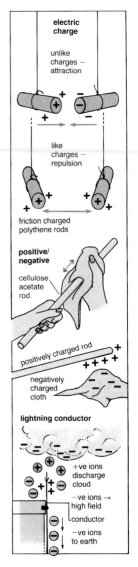

electric charge

unlike charges – attraction

like charges – repulsion

friction charged polythene rods

positive/negative

cellulose acetate rod

positively charged rod

negatively charged cloth

lightning conductor

+ve ions discharge cloud

−ve ions → high field

conductor

−ve ions to earth

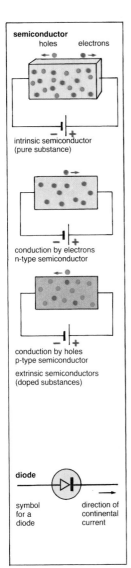

semiconductor
holes   electrons

intrinsic semiconductor
(pure substance)

conduction by electrons
n-type semiconductor

conduction by holes
p-type semiconductor

extrinsic semiconductors
(doped substances)

diode

symbol   direction of
for a   continental
diode   current

semiconductor (*n*) a substance or material with conducting properties halfway between those of a conductor (p.69) and an insulator (p.69). A pure substance used as a semiconductor, called an **intrinsic semiconductor**, conducts electric current by electrons and by **holes**. Holes are formed when an electron is removed from an atom in a crystalline structure. Holes can move from one atom to another under the influence of an electric potential, they act as carriers of a positive charge. Electrons act as carriers of negative charge, as in conductors. In an intrinsic semiconductor there are equal numbers of holes and electrons. The conducting power of a pure semiconductor is increased a thousand times by the addition of a very small amount of an impurity. Addition of an impurity is called **doping**, and the result is an **extrinsic semiconductor**. Some extrinsic semiconductors have more electrons than holes, these are **n-type semiconductors**. Others have more holes than electrons, these are **p-type semiconductors**. Examples are: intrinsic semiconductors — silicon, germanium; n-type semiconductor — silicon doped with phosphorus; p-type semiconductor — silicon doped with boron.

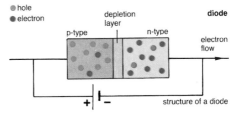

hole
electron

depletion
layer   **diode**

p-type   n-type

electron
flow

structure of a diode

diode[1] (*n*) a p-type semiconductor and an n-type semiconductor (↑) are joined at a junction. Holes and electrons diffuse across the junction, neutralize each other, and form a narrow **depletion layer** with no holes or electrons. If the p-type semiconductor is made positive and the n-type negative, the depletion layer is reduced and a current flows across the junction. If the voltage is reversed, the depletion layer is increased and very little, or no, current flows. The diode acts as a rectifier for alternating current. *See* **bias** (p.84).

transistor (*n*) a device consisting of two
semiconductor diodes (p. 83) joined together to
form an n-p-n (common) or p-n-p (less common)
structure. The transistor has three connections, the
emitter, the collector and the base.

emitter (*n*) the region in a transistor (↑) from which
current carriers flow, either electrons from n-type
regions or holes from p-type regions.

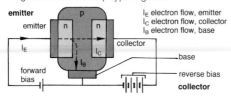

emitter

emitter

I_E

forward
bias

p

n   n

collector

I_B

I_C

base

I_E electron flow, emitter
I_C electron flow, collector
I_B electron flow, base

reverse bias

collector

collector (*n*) the region in a transistor into which
current carriers flow; it is the same type of
semiconductor (n- or p-) as the emitter (↑).

base[1] (*n*) a thin piece of semiconductor material
placed between the emitter and the collector (↑). It
is of a different type of semiconductor from the
emitter and collector. The voltage applied to the
base controls the current flow through the transistor.

bias (*n*) a voltage applied across a diode junction. If
the bias reduces the depletion layer of a diode
(p.83), it is called **forward bias** and it allows current
to pass. If the bias increases the depletion layer, it
prevents a current flowing, and is called **reverse
bias**.

n-p-n transistor the emitter and collector (↑) are
n-type semiconductors, the base is a p-type
semiconductor. A low forward bias (↑) is applied to
the base-emitter junction and a high reverse bias to
the base-collector junction. The electron flow into
the emitter, I_E, divides into I_B, the flow to the base,
and I_C, the flow to the collector. $I_E = I_B + I_C$, and I_C
≈ 100 I_B. The values for I_B and I_C for different values
of the voltage (V_{CE}) between collector and emitter
are shown in the graph. If a small varying current is
applied to the base in addition to the bias current,
then a much larger current flows from the collector
(taking current as an electron flow). The transistor
thus amplifies currents: p-n-p transistors function in
a similar fashion, but current is carried by holes,
and the sign of the bias is reversed.

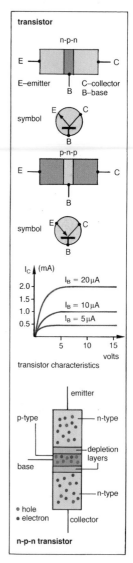

transistor

n-p-n

E   C

E–emitter   C–collector
B   B–base

symbol

E   C

B

p-n-p

E   C

B

symbol

E   C

B

I_C (mA)

2.0

1.5

1.0

0.5

$I_B = 20\,\mu A$

$I_B = 10\,\mu A$

$I_B = 5\,\mu A$

5   10   15
volts

transistor characteristics

emitter

p-type   n-type

base

depletion
layers

n-type

• hole
• electron   collector

n-p-n transistor

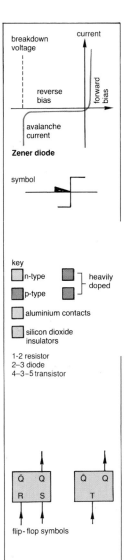

Zener diode

symbol

key

n-type
p-type

heavily doped

aluminium contacts

silicon dioxide insulators

1-2 resistor
2–3 diode
4–3–5 transistor

Q̄ Q
R S

Q̄ Q
T

flip-flop symbols

Zener diode a type of silicon junction diode (p.83). With reverse bias (↑) the transistor breaks down and a large current flows, which is independent of the voltage. Zener diodes can have breakdown voltages between 2.7 V and 200 V. The large current is called an **avalanche current**. Zener diodes are used to stabilize voltages in circuits.

integrated circuit the combination of a number of components, e.g. resistors, capacitors, transistors, into a complete circuit made from one piece of semiconductor material. The construction of such a device fron n-type and p-type semiconductors, silicon diode insulators and metal conductors is shown in the diagram. An integrated circuit is more reliable than a conventional circuit connected by wires, as there are no loose connections or short circuits.

integrated circuit

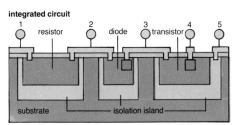

flip-flop (n) an electronic device which can have two stable states for an output, e.g. outputs of 0 volts or 5 volts. The input can have two terminals corresponding to the stable states, or one terminal which switches the device. The terminals are labelled R and S for two inputs, and T for one input. The output is labelled Q. A second output is labelled Q̄; it is always in the opposite state to Q but is not used. A pulse of current switches the flip-flop from one state to another, and the results are shown in the table.

two inputs

initial Q	0v	0v	5v	5v
input R	5v	—	5v	—
input S	—	5v	—	5v
final Q	0v	5v	0v	5v

one input

initial Q	0v	5v	0v	5v
input T	0v	0v	5v	5v
final Q	0v	5v	5v	0v

flip-flop

gate (*n*) an electronic switching circuit which has one or more inputs, but only one output. The output is energized for certain conditions of input. An AND gate has the output energized if all the inputs (two or more) are energized. An OR gate has the output energized if any one input (out of two or more) is energized. A NOT gate has one input and one output; the output is energized if the input is not energized, and vice versa.

photoelectric cell the cell consists of a curved cathode (p.75) and a small anode (p.75). The cathode is coated with material sensitive to light, e.g. selenium, caesium oxide. When light or ultraviolet radiation falls on the cathode, electrons are emitted. If a positive potential is applied to the anode, the electrons (p.64) are attracted to it, a circuit (p.72) is completed and current flows in the circuit. The photoelectric cell can be used to measure the intensity of a radiation. The device is also called a **photocell**.

photosensitive (*adj*) describes a substance which gives rise to a photoelectric effect when electromagnetic radiation falls on it.

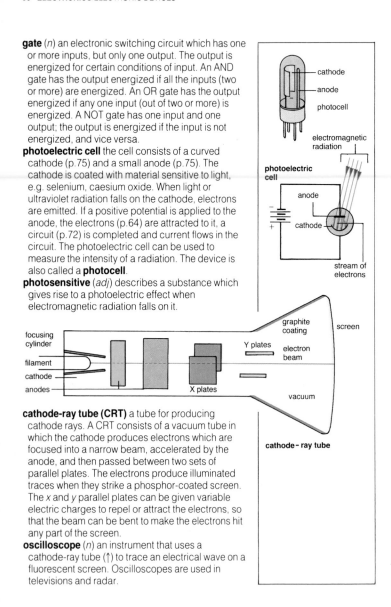

photocell

photoelectric cell

cathode-ray tube (CRT) a tube for producing cathode rays. A CRT consists of a vacuum tube in which the cathode produces electrons which are focused into a narrow beam, accelerated by the anode, and then passed between two sets of parallel plates. The electrons produce illuminated traces when they strike a phosphor-coated screen. The *x* and *y* parallel plates can be given variable electric charges to repel or attract the electrons, so that the beam can be bent to make the electrons hit any part of the screen.

oscilloscope (*n*) an instrument that uses a cathode-ray tube (↑) to trace an electrical wave on a fluorescent screen. Oscilloscopes are used in televisions and radar.

valve² (*n*) an electrical device through which an electric currrent passes in one direction only. Valves are classified according to the number of electrodes (↓) they contain, e.g. a diode contains two electrodes; a triode contains three electrodes.
thermionic valve an alternative name for a **valve** (↑).
electron tube an alternative name for a **valve** (↑).
tube (*n*) an alternative name for a **valve** (↑), used chiefly in America.
electrode² (*n*) a terminal that conducts a current into or out of an electrical device. The positive electrode is called the anode. The negative electrode is called the cathode.

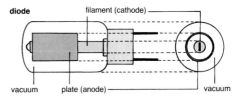

diode² (*n*) a valve with two electrodes: a positive electrode (anode) and a negative electrode (cathode). The cathode is heated to emit electrons which are attracted to the anode. There is thus a flow of negative charge from the negative to the positive electrode. The main application of diodes is for rectifying alternating current. Applying an alternating voltage to the anode and cathode produces current in one direction.

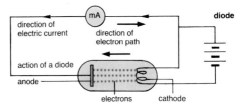

triode (*n*) a valve (↑) with three electrodes. It consists of a vacuum tube which contains an anode, a grid with a variable negative voltage, and a cathode. Voltage changes in the grid have a large effect on the electron current and give a big voltage at the anode. A triode is used to amplify voltage or power.

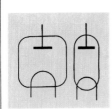

symbols used for a
diode

computer (*n*) an electrical machine which can both process and store information; the information may consist of numbers, or words, or both. A computer accepts information using an input unit. The information is processed by a CPU (p.90) or stored in memory (p.92) and processed afterwards. The result is supplied by an output unit. Instructions are usually given to the computer by a program (p.96).

calculator (*n*) a simple type of computer concerned only with numbers. Calculators vary in size from hand-held models to a full-sized computer. They are used for mathematical and arithmetical calculations that are supervised by an operator, as well as those which operate by means of a program.

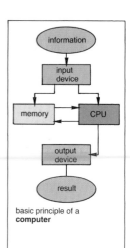

basic principle of a
computer

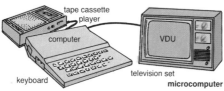

microcomputer

microcomputer (*n*) the smallest type of computer, usually designed for a single user. It consists of an input/output unit, a CPU (p.90), a memory (p.92) and a control unit (↓). Input is by means of a keyboard (↓). Output is given on a VDU (p.91), a tape cassette, or a floppy disk (p.94). The tape cassette and floppy disk can also be used for input. This forms a complete computing system, but less powerful than minicomputers (↓).

minicomputer (*n*) a computer larger than a microcomputer, capable of handling several input and output units. It can use more than one program language (↓).

mainframe computer the largest type of computer capable of storing large quantities of information and operating a network (p.97) of other computers and input/output units. Such computers are operated in banks and other large business firms.

word processor a specialized type of computer concerned only with words. It accepts statements in words, can check spelling, stores the words and can rearrange them for printing for such things as letters, reports and articles. It can use various styles of print, and can print many copies.

minicomputer

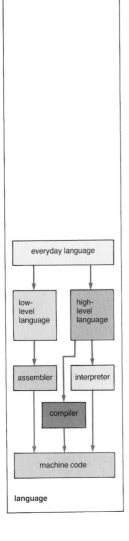

everyday language

low-level language

high-level language

assembler

interpreter

compiler

machine code

language

compute (*v*) to get an answer to a numerical problem by arithmetical methods. Originally this was the only function of computers. Improvements led to the use of formulae in computing and finally to the inclusion of words, as in tables and other displays of numerical facts.

hardware (*n*) the physical objects, metallic and plastic, that make a computer, or any device working with the computer, e.g. printer, floppy disk. The hardware is operated by software (p.96).

key (*n*) a marked button that can be pushed down to produce a computer-readable code for a letter, number, punctuation sign or mathematical sign. The common keyboard (↓) is called a QWERTY keyboard, with the keys arranged as on a typewriter.

keyboard (*n*) an input device that encodes data by pushing down a key (↑).

control (*n*) the parts of a computer which execute instructions given in a program. The instructions are held in memory (p.92), with a machine code (p.95) for each instruction, and are selected and interpreted before execution. The selection of instructions can depend upon specified conditions.

control unit a section of the CPU (p.90) that directs the operations for the execution of instructions. It sends command (↓) signals to other parts of the computer and to other output and memory devices which are working with the computer, e.g. a printer, floppy disk, etc.

command (*n*) an electric pulse which causes an operation to start, continue, or stop, or operates a gate (p.86). It is not the same as an instruction, which needs several commands, e.g. an instruction to a printer sends commands for ready, paper feed, start a motor, etc.

language (*n*) there are **low-level** and **high-level** languages. A low-level language is close to machine code (p.95) so that each instruction in a program (p.96) can be converted directly to machine code on a one-to-one basis. A high-level language consists of reserved words and symbols which are close to normal English words, e.g. PRINT, IF, LET, +, (). A statement in a high-level language program is converted to a low-level language and finally into machine code.

CPU (*abbr*) **central processing unit**. The unit in a computer that controls all other units. It consists of an arithmetic unit, a control unit (p.89), and a small internal memory for temporary storage. Buses (↓) are provided for data input and output and connections to memory (p.92).

bus (*n*) a path consisting of several conductors used for transmitting signals from one or more sources to one or more destinations. A bus is used to transfer data and commands inside a computer and between a computer and other devices working with a computer, e.g. printer, magnetic tape. Buses for data have 8 lines; a single wire is not considered to be a bus.

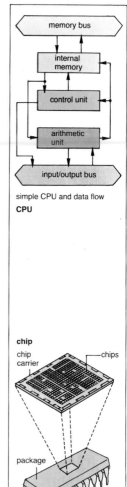

simple CPU and data flow

CPU

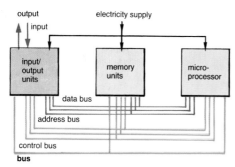

bus

microprocessor (*n*) the CPU (↑) of a microcomputer, usually available on a single silicon chip (↓). A dedicated microprocessor consists of a control unit and an arithmetic unit with a fixed set of instructions, so no internal memory is needed.

chip (*n*) a very small piece of a semiconductor, about 4–6mm square, on which microscopic electronic devices are formed as integrated circuits (p.85). The circuits are connected to external circuits by fine aluminium wires. The chip is hermetically sealed in a plastic case and mounted on a **chip carrier**. The chip carrier is mounted on a plastic or ceramic package, an oblong base. Metal pins provide contacts to external circuits and the whole is a chip component.

silicon chip a chip with silicon as the semiconductor; the chips in a microcomputer are usually silicon chips.

chip

wafer (*n*) a thin slice of silicon cut from a crystal of p-type silicon. Some wafers consist of n-type silicon, but p-type is generally used. A wafer is about 7.5–10 cm in diameter and circular or square in shape; it is about 0.45 mm thick. Individual silicon chips (↑) are cut from the wafer.

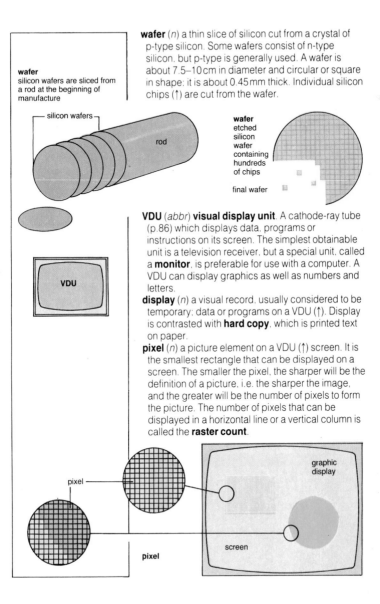

wafer
silicon wafers are sliced from a rod at the beginning of manufacture

silicon wafers

rod

wafer
etched silicon wafer containing hundreds of chips

final wafer

VDU

VDU (*abbr*) **visual display unit**. A cathode-ray tube (p.86) which displays data, programs or instructions on its screen. The simplest obtainable unit is a television receiver, but a special unit, called a **monitor**, is preferable for use with a computer. A VDU can display graphics as well as numbers and letters.

display (*n*) a visual record, usually considered to be temporary; data or programs on a VDU (↑). Display is contrasted with **hard copy**, which is printed text on paper.

pixel (*n*) a picture element on a VDU (↑) screen. It is the smallest rectangle that can be displayed on a screen. The smaller the pixel, the sharper will be the definition of a picture, i.e. the sharper the image, and the greater will be the number of pixels to form the picture. The number of pixels that can be displayed in a horizontal line or a vertical column is called the **raster count**.

pixel

pixel

graphic display

screen

LCD (*abbr*) **liquid crystal display**. A liquid is contained between two sheets of plastic. Tiny electrodes are placed in the liquid. When the electrodes are connected to an electrical supply, an electric field is formed between them. This field orients the liquid molecules; the molecules plane polarize light. The cover of the LCD also plane polarizes light, so that the two planes of polarization are at right angles and hence the liquid molecules appear black. The electrodes form seven bar-shaped segments which are used to display decimal numbers. When the electrical supply is switched off, the display disappears. LCDs use very little power.

memory (*n*) a device which stores information in the form off binary (p.95) coded data. A medium capable of recording and storing information. It usually refers to the main memory of a computer, that is, the memory used directly by the CPU (p.90). Data can be put into, and taken out of, a memory. Memory is also called **storage**. A **volatile** memory loses all its stored data when the power is switched off. A **non-volatile** memory retains all its data when the power is switched off, but data can be erased, or altered. A **permanent** memory retains all its data and cannot be erased or altered, whether the power supply is on or off.

RAM (*abbr*) **random access memory**. A memory with data stored in locations. It provides access to any location with the memory access time independent of the location. Each location holds one byte (p.95), and each location has a unique address. Data can be written into or read from any location. The number of locations is a multiple of K (1 K = 1024); a 16 K memory has 16 × 1024 = 16384 addressable locations.

ROM (*abbr*) **read only memory**. A memory whose contents are not intended to be altered; the contents are fixed during manufacture and form a permanent memory (↑). ROM is used to store permanent programs; data can be read from it, but cannot be altered. It stores small programs such as routines for arithmetical processes, control programs for operating various sections of a CPU, programs for the detailed steps of instructions to the computer, instructions for loading programs, etc.

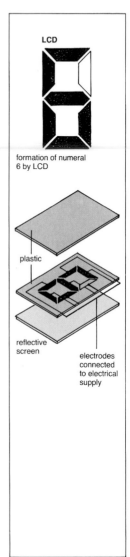

LCD

formation of numeral 6 by LCD

plastic

reflective screen

electrodes connected to electrical supply

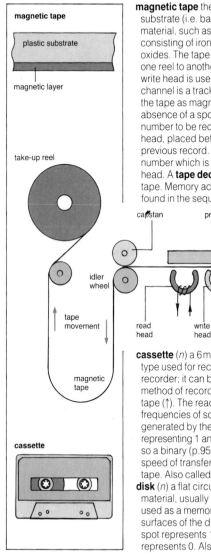

magnetic tape

plastic substrate

magnetic layer

take-up reel

capstan

idler wheel

tape movement

read head

write head

erase head

pressure pad

rewind capstan

tape from supply reel

magnetic tape

cassette

magnetic tape the tape consists of a plastic substrate (i.e. base) with a coating of magnetic material, such as ferrite, a chemical substance consisting of iron (III) oxide and other metallic oxides. The tape is stored on reels and moves from one reel to another, pulled by a capstan. A read and write head is used for each channel on the tape; a channel is a track on the tape. Data are recorded on the tape as magnetic spots to represent 1 and the absence of a spot to represent 0. This allows a number to be recorded in binary (p.95). An erase head, placed before the write head, removes any previous record. The write head records a binary number which is immediately checked by the read head. A **tape deck** controls the movement of the tape. Memory access is slow, as an item has to be found in the sequence of the tape.

cassette (*n*) a 6mm magnetic tape in a cassette, the type used for recording music and speech in a tape recorder; it can be used as a memory (↑). The method of recording differs from that of magnetic tape (↑). The read/write head reacts only to the frequencies of sound, so audio frequencies are generated by the computer with one frequency representing 1 and another frequency representing 0, so a binary (p.95) number can be recorded. The speed of transfer is much slower than for magnetic tape. Also called a **tape cassette**.

disk (*n*) a flat circular plate coated with a magnetic material, usually ferrite. *See* **magnetic tape**. It is used as a memory device. Data are stored on both surfaces of the disk by magnetic spots. A magnetic spot represents 1 in binary (p.95) and no spot represents 0. Also called a **magnetic disk**.

floppy disk a plastic disk (p.93) which is not rigid.
The disk is enclosed in a plastic case. Also called a
diskette. A disk can have 200 or more circular
tracks on each surface, and the magnetic spots for
recording data are recorded along a track. Each
disk track is divided into sectors. Any one sector or
any track can be addressed individually. The same
amount of data can be recorded on any track in a
sector, so the density of data is greater on inside
tracks than on outside ones. In **hard-sectored** disks
each sector is marked by a sector hole. In
soft-sectored disks, sector zero is marked by a
hole and subsequent sectors are identified by a
timing track. Floppy disks can be hard- or
soft-sectored. Read/write heads read or write data
on the tracks.

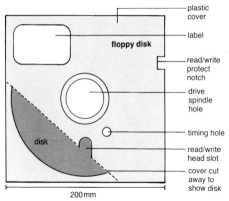

plastic cover
label
floppy disk
read/write protect notch
drive spindle hole
timing hole
disk
read/write head slot
cover cut away to show disk
200mm

hard disk a metal disk completely enclosed in an
airtight case to exclude dust. Several disks are
mounted on a single spindle and are all controlled
by the same disk drive. Hard disks are used only
with large computers. Hard disks have tracks and
sectors as for floppy disks (↑). The tracks on each
usable disk surface, with the same track number,
form a cylindrical slice down a stack of disks. The
cylinder has the same number as the track,
e.g. cylinder 86 consists of all tracks 86. A stack of 6
disks has 10 usable surfaces. Read/write heads
record and read data; all hard disks are
hard-sectored. *See* **floppy disk**.

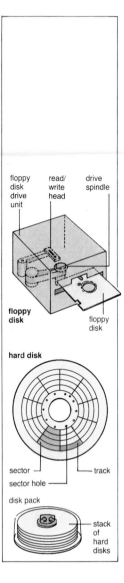

floppy disk drive unit
read/write head
drive spindle

floppy disk
floppy disk

hard disk

sector
sector hole
track

disk pack

stack of hard disks

code — ASCII code		
character	binary code	decimal
A	01000001	65
B	01000010	66
C	01000011	67
D	01000100	68
E	01000101	69
;	00111011	59
?	00111111	63
!	00100001	33
0	00110000	48
1	00110001	49
2	00110010	50
3	00110011	51

machine code

byte 1	byte 2	byte 3
00010010	00101000	10110110

byte 1

| 00010010 | add to CPU |

byte 2

| 00101000 | number in address |

byte 3

| 10110110 | 182 |

number in address 182 is 687, and this is added to arithmetic unit in CPU

code (*n*) a system of letters, numbers and symbols for converting one form of information to another. Also the representation of information using such a system. A computer uses several different codes in its operation, e.g. ASCII code for numbers, letters and punctuation marks, and machine code (↓) for computer commands. In ASCII code, the numbers 0 to 255 are used to represent letters, numerals, punctuation marks, mathematical symbols, e.g. 65 represents A, 83 represents S, and 67 represents C; these are capital letters, other numbers are used to represent lower case letters.

machine code a command in code that instructs a computer to perform a particular operation when reading a particular symbol. The complete machine code defines all the operations a computer can do. Each computer has a machine code suited to its CPU (p.90). Machine code is in binary (↓); all computers recognize only binary code. The binary code is organized in bytes (↓).

binary (*n*) a method of representing a number using only the numerals 1 and 0. In binary the number 10 represents decimal number 2. *See table* for correspondence between binary and decimal numbers.

digit (*n*) an arabic numeral with a place value in a number, e.g. 312 is a number with three digits; 444 is a number with three different digits but only one type of numeral. **digital** (*adj*).

bit (*n*) an abbreviation for binary (↑) digit (↑); it has a value of either 1 or 0. It is the smallest bit of information, and it is equal to one binary decision, i.e. true/false; yes/no; available/not available. A bit is both a digit in binary, and also the physical representation of a binary digit, i.e. a pulse of current/no pulse; magnetized spot/no magnetized spot; a hole in a card/no hole.

byte (*n*) a group of 8 bits considered as a single unit.

nibble (*n*) a group of 4 bits considered as a single unit. Two nibbles form a byte.

megabyte (*n*) describes the capacity of a memory which holds one million bytes (↑).

bug (*n*) an error, mistake, or a defect in a program; any fault in a computer or an electronic circuit causing a malfunction. The removal of bugs is called debugging.

surge (*n*) a short-term increase in voltage from a power supply. It can be high enough to damage electronic components. A surge can last for several cycles of an alternating current, or up to several hundred cycles.

data (*n*) any group of facts, numbers, symbols or letters which describe a state, value, or condition. Data usually describes a fact or an event. Data is contrasted with information, which is conveyed by a set of data, and with program, which is a set of instructions operating on data.

data processing all the operations carried out on data (↑) by means of a program in order to produce information or a specific result. It includes the rearrangement of data into a suitable form for further use.

data base a large and continuously updated collection of data stored in memory in such a way as to allow access to various parts of the data quickly and economically. By using subject headings, key phrases, etc, users can search for data, then sort, analyze and print-out information as required. The advantages of a data base are: always up-to-date; no duplicated information; ease of organizing new applications; easier file protection. The disadvantages are the need of a large software (↓) system and a fairly large computer; exposure to failure of the memory system.

software (*n*) the programs (↓), including systems of operation, procedures, standard routines on processing files; the solution of specific problems such as a payroll application for a firm; a group of routines is called a package; all are written for a particular computer system and supplied by the hardware (p.89) manufacturer. It does not include programs written by the user.

flowchart (*n*) a representation of a sequence of operations by conventional symbols showing the flow of information in a solution to a problem. The relationships between the different types of operations are shown in a series of steps; this helps a programmer to design an effective program.

program (*n*) a set of instructions, arranged in sequence, prepared for the direction of a computer to perform the necessary operations for the solution of a problem or the completion of a task. A program

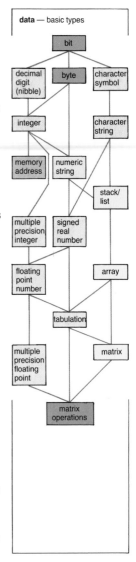

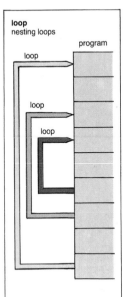

loop
nesting loops

program
loop
loop
loop

is usually developed from a flowchart (↑). The program is written in a high-level language, which is usually BASIC for a microcomputer.

loop (*n*) a segment of a program (↑) which repeats itself either a given number of times or until a required condition is met. Loops can be nested, which means that one loop is enclosed within another loop. An inner loop is completed for each cycle of an outer loop.

simulator (*n*) a program (↑) which represents a situation on a computer (p.88) system such that the decisions and actions of the user are the same as if he were actually in that situation, e.g. learning to fly an aeroplane. This use of a computer helps in situations which would otherwise be too costly. A simulator is also a program which executes a machine code (p.95) program on a computer when the machine code is designed for a different computer. This tests the program before it is put into ROM (p.92).

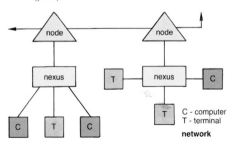

node

node

nexus

T   nexus   C

C   T   C

T   C - computer
T - terminal

network

network (*n*) an arrangement of computers and terminals by interconnection. A terminal generally has input and output operations only. Terminals can be local or remote. Local connections are by cable, remote connections are by telecommunication. The network function of linking units is mainly performed by the hardware (p.89) of the network units. Networks have a **nexus** which controls a group of units and a **node** which links to nexuses in one direction and to another node in the opposite direction.

channel (*n*) a physical path along which information can be transmitted, e.g. the path connecting units in a network (↑).

radiation (*n*) the process by which energy is sent out in the form of electromagnetic waves, i.e. infrared light, visible light, ultraviolet, radio and radar waves, gamma rays and X-rays.

intensity (*n*) the quantity of energy carried by a wave motion in 1 second through an area of $1 m^2$ perpendicular to the direction of travel. The intensity of light is its brightness; the intensity of sound is its loudness. The intensity decreases proportionally to the square of the distance from the source in a dense medium.

light (*n*) a type of radiation detected by the eye. It is propagated as a wave motion with a speed of $3 \times 10^8 ms^{-1}$. Its wavelengths are between $4 \times 10^{-7}m$ and $7 \times 10^{-7}m$. White light can be split into a spectrum of red, orange, yellow, green, blue, indigo and violet light.

infrared (i.r.) rays invisible electromagnetic waves which can be felt as heat. Their wavelength is just longer than that of red light, which we can see. They can pass through mist and cloud without being scattered.

ultraviolet (u.v.) rays invisible electromagnetic waves with a wavelength that is slightly shorter than that of visible violet light. Ultraviolet rays act on photographic film and plates.

discharge tube an evacuated glass bulb or tube in which a continuous stream of electrons pass from the cathode to the anode due to a high applied potential.

cathode rays the radiation which is produced by a cathode (p.75) in a discharge tube (↑). The rays consist of a stream of electrons.

X-rays (*n*) invisible electromagnetic waves with a very short wavelength. X-rays pass through many substances that stop ordinary light rays. The absorption of the rays depends in part on the density of a substance, e.g. bone absorbs more X-rays than skin and muscle — a fact utilized in X-ray photographs.

X-ray tube a tube for producing X-rays (↑). It consists of a heated filament (the cathode) that emits a stream of electrons which are focused by a cup onto a copper block anode called a **target**. The anode emits X-rays when it is bombarded by the high-energy electrons.

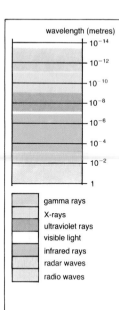

radiation detected by the eye.

gamma rays
X-rays
ultraviolet rays
visible light
infrared rays
radar waves
radio waves

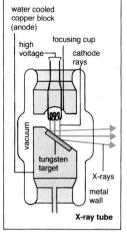

X-ray tube

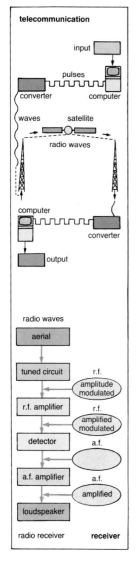

telecommunication

radio receiver **receiver**

telecommunication (*n*) the communication of information by electrical signals along wires or by radio circuits using radio waves (p.101). The original information can be alphabetic, numeric, pictorial, or it can represent a measurement. The types of telecommunication are telegraphy, radio, television and facsimile (p.102).

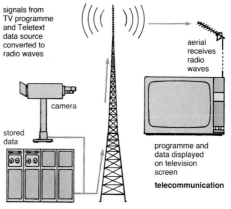

signals from TV programme and Teletext data source converted to radio waves

camera

stored data

aerial receives radio waves

programme and data displayed on television screen

telecommunication

transmit (*v*) to pass on energy from one place to another through a medium. Transmit is used incorrectly to send out electrical signals by radio waves.

transmitter (*n*) an electrical device which sends out a radio wave (p.101) which carries a signal by modulation (p.100) of the wave. A transmitter is used for radio telegraphy, radio telephony, broadcasting and television.

antenna (*n. pl. antennae*) a wire, or a wire structure, for sending out or receiving radio waves (p.101).

aerial[1] (*n*) alternative word for antenna (↑).

receiver (*n*) an electrical device for receiving radio waves (p.101). An antenna (↑) circuit is tuned to the radio wave required for reception. The radio wave is demodulated (p.100) by rectification, using a diode or a transistor, and this produces audio frequencies. The audio frequencies are amplified by an amplifying circuit and passed to a loudspeaker (p.78) or an earphone. A receiver is also called a radio set.

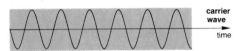

carrier
wave

time

carrier wave a radio wave (↓) of constant frequency
and constant amplitude given out by a radio
transmitter. Signals are superimposed on the carrier
wave by modulation (↓).

modulation (*n*) the wave form of an audio frequency
signal alters the amplitude or the frequency of the
carrier wave (↑). In amplitude modulation the
amplitude of the carrier wave is altered to the
waveform of the signal, but the frequency remains
constant. In frequency modulation, the amplitude
remains constant and the frequency varies above
and below its nominal value. **modulate** (*v*).

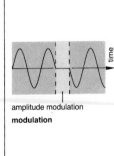

time

amplitude modulation
modulation

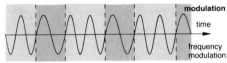

modulation

time

frequency
modulation

demodulation (*n*) the separation of audio frequency
waves from radio waves by a demodulator. The
radio wave is rectified and passed through a filter
(p.81) which rejects the high frequency radio wave
and transmits the low frequency audio wave.
demodulate (*v*).

signal (*n*) a sign used to convey information; light,
sound, electric current, motion of an object, can all
be used to make a sign. In a computer, signals are
pulses of current giving coded data. In a radio set,
signals are superimposed on carrier waves by
modulation (↑). Radio waves are also used to
transmit video (p.102) signals. In telegraphy,
signals are sent in Morse code (p.79).

distortion[2] (*n*) unwanted change in the waveform of
a signal (↑), usually caused by amplification in a
circuit not being a direct proportion of the input
signal.

noise (*n*) electrical disturbances of various origins
causing false signals (↑); the random variation in
current, voltage, signal strength associated with
any electrical circuit working in normal
surroundings. False signals from noise can
introduce errors into data in computers.

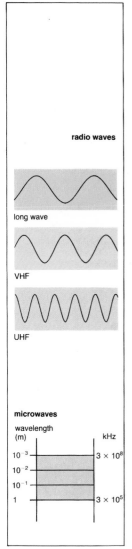

kHz	waves		wavelength (m)
3×10^8	infrared waves		10^{-3}
	radar	microwaves	10^{-2}
			10^{-1}
3×10^5			1
30000	VHF	short waves	10
3000	HF		10^2
300	MF	medium waves	10^3
30	LF	long waves	10^4
3	VLF	very low frequencies	10^5
	alternating current		
0.05			

radio waves

long wave

VHF

UHF

microwaves

wavelength (m)   kHz

10^{-3} —— 3×10^8
10^{-2} ——
10^{-1} ——
1 —— 3×10^5

radio (n) the use of certain electromagnetic waves (radio waves) to transmit electrical signals without wires. The waves are transmitted from an aerial, and picked up by a receiving aerial that feeds a receiver.
radio wave an electromagnetic wave with a frequency between 3 kHz and 3×10^8 kHz. Alternating currents with frequencies of 50 Hz are also included in the radio frequencies of electromagnetic waves.
long wave a radio wave (↑) with a wavelength between 1 km and 10 km, corresponding to a frequency between 300 and 30 kHz; also called low frequency (LF) waves. This range forms the long waveband, used in local broadcasting, but used less than the medium waveband (↓).
medium wave a radio wave (↑) with a wavelength between 100 m and 1 km, corresponding to a frequency between 3000 and 300 kHz; also called medium frequency (MF) waves. This range forms the medium waveband, used in local broadcasting.
short wave a radio wave (↑) with a wavelength between 1 m and 100 m. It is divided into two wavebands, high frequency (HF) with a frequency between 30000 and 3000 kHz, and very high frequency (VHF) with a frequency between 300000 kHz and 30000 kHz. Short waves are used for long distance broadcasting.
microwave (n) a radio wave (↑) with a frequency between 3×10^8 kHz and 300000 kHz. These frequencies are highly directional and are used in radar (p.103). They also extend into infrared waves.
audio frequency a frequency of a sound wave between 30 Hz and 20000 Hz, the limits of audibility for the human ear.

teleprinter (*n*) a device like a typewriter which transmits or receives printed messages by telegraph line. Each letter, numeral, punctuation mark, is coded in binary and the machine encodes or decodes the message.

facsimile (*n*) pictorial information is scanned electronically and converted to electrical signals. The screen is divided into pixels (p.91) and the electrical signal indicates the shade of white, grey or black of each pixel. This produces a picture similar to that seen in a newspaper. The receiver of the signals, besides displaying the picture, can also print the picture.

fax (*abbr*) **facsimile** (↑).

television (*n*) a system for sending and receiving sound and pictures using radio waves (p.101). A television camera converts the light waves into electrical impulses. They are sent to a television receiver (p.99), where they are converted into a picture on the screen of a cathode-ray tube (p.86) in the receiver.

scan (*v*) the sampling in sequence of the conditions or physical state of each separate unit of a system, e.g. sampling in sequence each pixel (p.91) on a television screen.

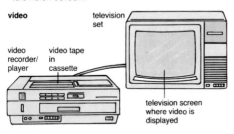

video

television set

video recorder/ player

video tape in cassette

television screen where video is displayed

video (*n*) the display on a television screen; the conversion of electrical impulses to a visual signal. Also describes frequencies similar to those used in television screening.

VCR (*abbr*) **video cassette recorder**. Electrical signals from a television set, for both sound and picture, are recorded on magnetic tape. Different tracks on the magnetic tape are used for audio frequencies and video frequencies; a control track is also used to match sound and picture. A

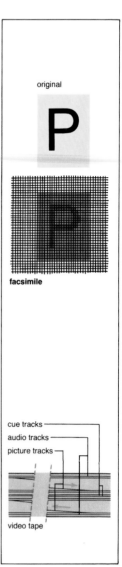

original

facsimile

cue tracks
audio tracks
picture tracks

video tape

read/write head is used, so the tape can be played back, and the picture then appears on the television screen and is accompanied by sound.

VTR (*abbr*) **video tape recorder**. Video cassette recorder (↑).

record (*v*) to make a magnetic, photographic or mechanical representation of a signal on a suitable medium, e.g. to record audio frequencies on magnetic tape.

recording head a read/write head used with magnetic tape in tape decks, cassettes (p.93) and VCRs (↑). *See* **magnetic tape** (p.93) and **cassette**.

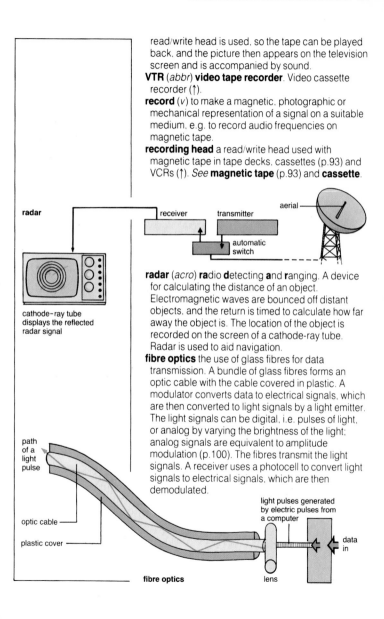

radar

cathode-ray tube
displays the reflected
radar signal

radar (*acro*) **ra**dio **d**etecting **a**nd **r**anging. A device for calculating the distance of an object. Electromagnetic waves are bounced off distant objects, and the return is timed to calculate how far away the object is. The location of the object is recorded on the screen of a cathode-ray tube. Radar is used to aid navigation.

fibre optics the use of glass fibres for data transmission. A bundle of glass fibres forms an optic cable with the cable covered in plastic. A modulator converts data to electrical signals, which are then converted to light signals by a light emitter. The light signals can be digital, i.e. pulses of light, or analog by varying the brightness of the light; analog signals are equivalent to amplitude modulation (p.100). The fibres transmit the light signals. A receiver uses a photocell to convert light signals to electrical signals, which are then demodulated.

path
of a
light
pulse

optic cable

plastic cover

fibre optics

test (*n*) (1) an experiment to determine whether a particular substance is present, and what its properties are. (2) a trial run to determine whether a machine or instrument is working correctly. **test** (*v*).

observation (*n*) the use of human physical senses, in conjunction with previous knowledge and experience, to note facts or events, e.g. the obsevation that blue litmus paper turns red when placed in an acid. **observe** (*v*).

identification (*n*) the recognition and naming of a substance by recognizing its known properties; the recognition and naming of a process or form of energy by comparing its characteristics with those of a known process or form of energy. **identify** (*v*).

sample (*n*) a small quantity taken from a large quantity in order to investigate the properties of a substance or material, e.g. in the manufacture of pure substances, a sample is taken at regular intervals to test the purity of the substance.

analyze (*v*) to find the constituents of a mixture or compound; to find the concentration of solute in a solution; to find the percentage composition by mass of a compound. **analysis** (*n*).

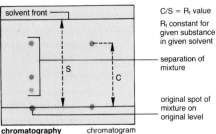

C/S = R$_f$ value

R$_f$ constant for given substance in given solvent

separation of mixture

original spot of mixture on original level

chromatography   chromatogram

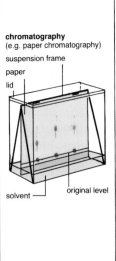

chromatography
(e.g. paper chromatography)

suspension frame

paper

lid

original level

solvent

chromatography (*n*) a method of separating a mixture of solutes by the use of a porous material. In paper chromatography a mixture of substances is dissolved in a suitable solvent and spots of the solution put on a strip of prepared paper (similar to filter paper). The paper is suspended in a closed tank with the paper dipping into a solvent. The solvent ascends the paper. The solvent travels furthest, while the solutes from the mixture ascend to different heights, thus separating the different substances for analysis.

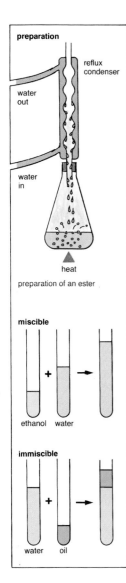

preparation

reflux condenser

water out

water in

heat

preparation of an ester

miscible

ethanol   water

immiscible

water   oil

preparation (*n*) (1) the stages of getting ready the apparatus and substances to be used in an experiment. (2) a substance obtained under laboratory conditions for a definite purpose, e.g. a specimen of hydrogen obtained by experiment is a preparation.

identical (*adj*) describes two or more substances, processes or radiations which are exactly alike, i.e. they have exactly the same properties and characteristics, and exactly the same number of properties and characteristics.

similar (*adj*) describes two or more substances, processes or radiations which have many properties or characteristics which are exactly alike, but have a few which are different. **similarity** (*n*).

physical change a change that does not alter the chemical composition of a substance, e.g. a metal expanding on heating; water vapour condensing to form water droplets; salt dissolving in water.

chemical change a change of one substance into another, different substance; the new substance has a different chemical composition and possesses different chemical properties. Copper and sulphur undergo a chemical change when they react to become copper sulphide.

combination (*n*) the joining of two or more elements or compounds to form a new substance, e.g. the combination of hydrogen and oxygen to form a molecule of water; the combination of zinc and bromine to form zinc bromide.

formation (*n*) (1) bringing about an effect by a physical change (↑), e.g. the formation of ice by cooling water below its freezing point. (2) bringing about the presence of a substance by a chemical change (↑), e.g. the formation of sulphur dioxide by adding an acid to a sulphite.

miscible (*adj*) describes liquids which will mix together completely, e.g. ethanol and water are miscible.

immiscible (*adj*) describes liquids which will not mix together, they will separate into layers, e.g. oil and water are immiscible.

separate (*v*) (1) to divide a mixture into its constituent parts, or to obtain one of those parts. (2) of immiscible (↑) liquids, to come apart and form layers. **separation** (*n*), **separable** (*adj*).

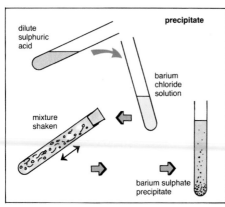

dilute sulphuric acid

precipitate

barium chloride solution

mixture shaken

barium sulphate precipitate

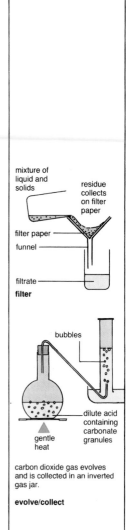

mixture of liquid and solids

residue collects on filter paper

filter paper

funnel

filtrate

filter

bubbles

dilute acid containing carbonate granules

gentle heat

carbon dioxide gas evolves and is collected in an inverted gas jar.

evolve/collect

precipitate (*n*) an insoluble substance made to appear by the addition of one solution to another solution, or by passing a gas through a solution, e.g. when a solution of sulphuric acid is added to a solution of barium chloride a precipitate (barium sulphate) is formed.

filter[2] (*v*) to separate an insoluble solid from a mixture of a liquid and a solid. e.g. to filter off solid impurities from a solution. **filter** (*n*).

filtrate (*n*) the solution which has been passed through a filter.

residue (*n*) the solid material held back when a mixture of a solid and a liquid passes through a filter.

evolve (*v*) to form and give off a continuous stream of bubbles of gas due to a chemical reaction, e.g. when a dilute acid is added to a carbonate, carbon dioxide gas is evolved.

collect (*v*) in a chemical reaction, to separate (p.105) and gather a constituent part of a product, e.g. to collect a gas over water.

bubble (*n*) (1) a sphere of gas either rising through a liquid or enclosed within a thin liquid film. (2) (*v*) to make a stream of bubbles of a gas pass through a liquid, e.g. to bubble air through sodium hydroxide solution to remove carbon dioxide gas.

sublime (*v*) to change directly from a solid to a vapour, and then from a vapour to a solid, without the formation of a liquid.

when exposed to air, lumps of crystalline e.g. washing soda lose water of crystallization and turn to powder

water lost

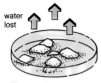

efflorescent

oxidation

electron lost

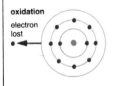

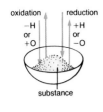

oxidation | reduction
−H or +O | +H or −O

substance

electron gained

reduction

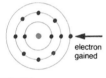

volatile (*adj*) a liquid that readily changes to a gas (p.42) or vapour (p.43), e.g. chloroform is a volatile liquid. **volatility** (*n*).

odour (*n*) that quality of a substance that is recognized by the sense of smell, e.g. the odour of vinegar.

odourless (*adj*) describes a material or substance that is totally without odour (↑), e.g. oxygen is an odourless gas.

deliquescent (*adj*) describes a substance that can pick up so much moisture from the air that it dissolves in it and becomes a solution. **deliquescence** (*n*).

hygroscopic (*adj*) describes a substance that tends to absorb and retain moisture from the air. If left in the atmosphere it will become damp.

efflorescent (*adj*) describes crystals (p.48) which tend to lose their water of crystallization (p.48) when exposed to air, and may eventually decompose (p.45) to a powder. **efflorescence** (*n*). **effloresce** (*v*).

reagent (*n*) a substance that causes a known and definite reaction with a particular chemical. Reagents are used to detect substances or to bring about a particular chemical reaction, e.g. bromine is a reagent which causes an addition reaction when added to ethene.

agent (*n*) a substance or force that is active in bringing about a particular chemical or physical effect, e.g. light is the agent that brings about photosynthesis (p.223).

oxidation (*n*) (1) the process by which a substance gains oxygen or loses hydrogen, e.g. the oxidation of sulphur to sulphur dioxide; the oxidation of hydrogen bromide to bromine. (2) the process by which electrons are lost from an ion (p.76) or atom, e.g. the oxidation of iron (II) to iron (III). **oxidize** (*v*).

reduction (*n*) the process by which a substance gains hydrogen or loses oxygen, e.g. the reduction of zinc (II) oxide to zinc; the reduction of bromine to hydrogen bromide. (2) the process by which electrons are gained by an ion or atom, e.g the reduction of iron (III) to iron (II).

corrosion (*n*) the gradual wearing away of the surface of a metal caused by chemical action, e.g. the corrosion of iron due to moisture and air.

corrosive (*adj*) describes any substance that causes corrosion of a metal, or attacks animal tissue by chemical action.

protected from atmosphere   corroded by atmosphere

rust (*n*) the corrosion (p.107) of iron; it is the reddish-brown coat of oxide that forms on iron exposed to the atmosphere. **rust** (*v*).

tarnish (*v*) the process whereby the surface of a bright metal becomes dull when exposed to the air. It is due to the corrosion (p.107) of the metal.

bleach (*v*) to make an object become white, or lose colour by chemical action or by exposure to the rays of the Sun.

thermal dissociation the temporary and reversible breakdown of a substance into smaller, simpler substances when it is heated. When the temperature falls these substances recombine to form the original substance, e.g. the thermal dissociation of ammonium chloride produces ammonia and hydrogen chloride gases; the two gases combine on cooling to form ammonium chloride.

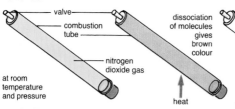

pyrolysis (*n*) the irreversible chemical decomposition of a substance into simpler substances, caused by heat. **pyrolytic** (*adj*).

violent (*adj*) describes a chemical reaction which is particularly vigorous or forceful, and during the course of which a large quantity of energy is suddenly released.

catalyst (*n*) a substance that alters the rate of a chemical reaction without itself being changed.

catalysis (*n*) the changing of the rate of a chemical reaction by the addition of a catalyst. **catalyze** (*v*). **catalytic** (*adj*).

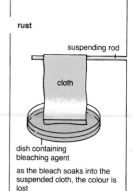

rust

suspending rod

cloth

dish containing bleaching agent

as the bleach soaks into the suspended cloth, the colour is lost

bleach

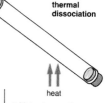

thermal dissociation

heat

at higher temperatures nitrogen dioxide dissociates to nitric oxide and oxygen which is colourless. The gas associates on cooling

impurity (*n*) a small amount of an unwanted foreign substance found in another substance, e.g. arsenic is often found as an impurity in metal ores.
purify (*v*) to rid a substance of its impurities.
air (*n*) the mixture of gases surrounding the Earth that make up its atmosphere. Air is invisible and odourless. It is comprised of approximately 78% nitrogen, 21% oxygen, 0.03% carbon dioxide, and the remainder consists of rare gases such as neon, helium and argon, and some water vapour.

composition of dry
air

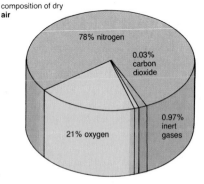

78% nitrogen

0.03% carbon dioxide

0.97% inert gases

21% oxygen

heat and light given out
combustion of gas

pipe

jet — unburnt gas

gas supply

chemical reaction between gas and oxygen in air

flame

combustion (*n*) a chemical reaction in which a substance combines with oxygen and gives out heat and light, and in some cases forms flames.
Rapid combustion produces a great deal of heat and light; **slow combustion** produces relatively little heat and no flame.
flame (*n*) a mass of hot vapour or gas that burns with a bright light, and gives out heat by combustion.
explosion (*n*) a violent (↑) chemical reaction caused by the sudden expansion of gases produced by rapid combustion (↑). **explode** (*v*), **explosive** (*adj*).

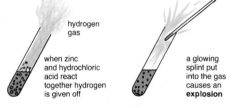

hydrogen gas

when zinc and hydrochloric acid react together hydrogen is given off

a glowing splint put into the gas causes an **explosion**

acid (*n*) a substance that forms hydrogen ions (p.76) in solution; it contains hydrogen which can be replaced by a metal, or by a base (↓), to form a salt. Many acids are corrosive (p.108) and most will change the colour of an indicator. **acidic** (*adj*), **acidify** (*v*).

base² (*n*) a substance, generally a metal oxide or metal hydroxide, which reacts with an acid to produce a salt (↓) and water only. Many bases are insoluble. **basic** (*adj*).

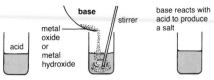

base reacts with acid to produce a salt

neutralization (*n*) a chemical reaction between an acid and a base in which both are destroyed in the formation of a salt, e.g. sodium chloride is formed when sodium hydroxide neutralizes hydrochloric acid.

alkali (*n*) a soluble base (↑) which in solution forms hydroxyl −OH (↓) ions.

hydroxyl (*n*) the atomic group −OH; occurs in solutions as the negatively charged hydroxyl ion OH− with an electrovalency of −1.

amphoteric (*adj*) having both acidic and basic properties, e.g. aluminium oxide neutralizes acids to form aluminium salts and water and it also neutralizes bases to form aluminates.

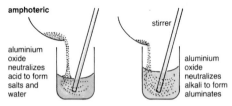

aluminium oxide neutralizes acid to form salts and water

aluminium oxide neutralizes alkali to form aluminates

salt (*n*) a compound formed when a metal replaces all or some of the hydrogen of an acid. The salt is named from the metal and the acid, e.g. copper sulphate. Soluble salts in solution form ions (p.76), a cation, e.g. Cu^{2+}, from the metal and an anion, e.g. SO_4^{2-} from the acid.

dilute hydrochloric acid is run slowly from the burette to the flask of sodium hydroxide solution; the acid has neutralized the alkali at the point of change in the indicator colour

acid

flask is swirled

salt being formed

base

flask contains an indicator to show when the neutral point is reached

neutralization

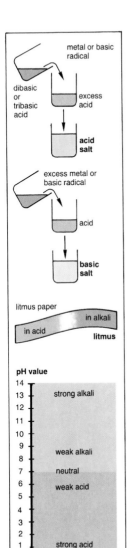

acid salt a salt formed when there is not enough base to neutralize (↑) an acid completely. They are formed with dibasic (↓) acids, when only part of the replaceable hydrogen in the acid is replaced by a metal, e.g. sodium hydrogen carbonate.

basic salt a salt (↑) formed when there is not enough acid to neutralize (↑) a base completely. It consists of a normal salt (↑) combined with a definite proportion of the base, e.g. basic lead carbonate, $2PbCO_3.Pb(OH)_2$.

basicity (*n*) (1) the number of hydrogen atoms (p.45) which can be replaced by a metal (p.44) in one molecule (p.45) of an acid, e.g. sulphuric acid H_2SO_4 has two replaceable hydrogen atoms and therefore a basicity of two. (2) the number of hydrogen ions (p.76) formed from one molecule of an acid.

monobasic (*adj*) describes an acid having a basicity of one, e.g. hydrochloric acid.

dibasic (*adj*) describes an acid having a basicity of two, e.g. sulphuric acid. Dibasic acids are able to form acid salts.

tribasic (*adj*) describes an acid having a basicity of three, e.g. phosphoric acid. A tribasic acid can form three series of salts.

indicator (*n*) a chemical, or mixture of chemicals, that changes colour in different environments. Indicators are used to show the presence of a particular substance, or the stage of a chemical reaction, or whether a solution is acidic or alkaline. **indicate** (*v*).

litmus (*n*) a purple indicator (↑) that turns blue in alkalis and red in acids. It is used to indicate neutralization (↑). When litmus is added to an alkali it turns blue. Acid is added to the alkali and when the litmus is just turning red, the neutralization is complete.

litmus paper absorbent paper that has been soaked in a litmus (↑) solution; it is used to indicate whether a solution is acidic or alkaline.

pH value an indiction of the acidity or alkalinity of a solution, based on the concentration of hydrogen ions (p.76); pH values range from 0 to 14. Between 0 and 7 a solution is acidic, and between 7 and 14 a solution is alkaline. A pH value of 7 is considered to be neutral.

radical (*n*) a group of chemically bonded atoms, found in compounds, that participate in a chemical reaction but remain unchanged. Radicals are unable to exist independently.

$$MgSO_4 + 2NaOH \rightarrow NaSO_4 + Mg(OH)_2$$
radical

inorganic (*adj*) describes elements (p.44) or compounds (p.45) that are derived from mineral sources. All elements and compounds are inorganic, apart from those containing carbon. However, carbonates (↓) and the oxides of carbon are considered as inorganic compounds.

oxide (*n*) a compound of oxygen and one other element, e.g. magnesium oxide, MgO, hydrogen oxide, H_2O.

peroxide (*n*) an oxide that reacts with cold dilute sulphuric acid to yield hydrogen peroxide H_2O_2, e.g. sodium peroxide Na_2O_2, the normal oxide is Na_2O.

hydroxide (*n*) a compound containing the hydroxyl radical (↑) −OH. Hydroxides are formed by the replacement of one of the hydrogen atoms in a molecule of water by some other atom or group, e.g. sodium hydroxide NaOH. Soluble hydroxides in solution form hydroxyl ions.

chloride (*n*) a salt (p.110) of hydrochloric acid. A chloride is also considered to be a compound consisting of chlorine and one other element, e.g. nitrogen trichloride NCl_3.

sulphate (*n*) a salt of sulphuric acid, H_2SO_4, containing the sulphate radical (↑), e.g. calcium sulphate $CaSo_4$.

sulphite (*n*) a salt of sulphurous acid, H_2SO_3, containing the sulphite radical, e.g. calcium sulphite $CaSo_3$.

nitrate (*n*) a salt of nitric acid, HNo_3, containing the nitrate radical, e.g. calcium nitrate $CaNo_3$.

nitrite (*n*) a salt of nitrous acid, HNo_2, e.g. calcium nitrite $CaNo_2$.

carbonate (*n*) a salt (p.110) formed by a metal (p.44) and carbon dioxide in solution, containing the radical Co_3 from carbonic acid, H_2Co_3, e.g. calcium carbonate $CaCo_3$.

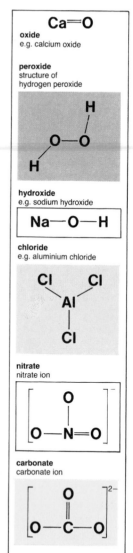

oxide
e.g. calcium oxide

peroxide
structure of
hydrogen peroxide

hydroxide
e.g. sodium hydroxide

chloride
e.g. aluminium chloride

nitrate
nitrate ion

carbonate
carbonate ion

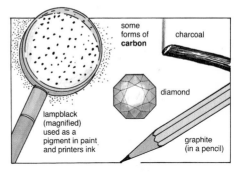

some forms of **carbon**

charcoal

diamond

lampblack (magnified) used as a pigment in paint and printers ink

graphite (in a pencil)

hydrocarbon
e.g. methane – the main constituent of natural gas

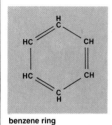

benzene ring

carbon (*n*) a non-metallic element with three allotropic (p.44) forms; it has an atomic number of 6 and a relative atomic mass of 12.01. Carbon atoms have a covalency (p.65) of 4, and are able to unite with each other to form large and varied compounds, many of which are found in animals and plants. The chemical symbol for carbon is C.

organic (*adj*) refers to materials, substances and compounds containing molecules (p.45) with chains or rings of carbon atoms, though not including the oxides of carbons and the carbonates. All living organisms consist of organic compounds.

hydrocarbon (*n*) an organic compound containing only hydrogen and carbon atoms. Hydrocarbons may be structured in the form of a ring of carbon atoms with hydrogen combined, or they may be comprised of a long chain of carbon atoms joined by covalent bonds to hydrogen atoms. Petroleum and natural gas are hydrocarbons.

benzene ring the structure of a benzene molecule; it consists of six carbon atoms joined by covalent bonds in the shape of a hexagonal ring, with alternate double and single bonds between the carbon atoms in the molecule. The benzene ring also occurs in other compounds in which rings are fused together.

oil (*n*) any of a number of greasy, flammable substances that are normally liquid at room temperature. Oils are obtained from certain plants and animals and from mineral deposits. The esters of glycerol and organic (↑) acids form the oils and fats (p.165) in living organisms. **oily** (*adj*).

distillation (*n*) the process of separating a mixture of different liquids, based on the differences in their boiling points. As the mixture is heated the liquids boil off in order of their increasing boiling points; the vapour from each liquid is condensed (p.41) and then collected. **distil** (*v*), **distillate** (*n*).

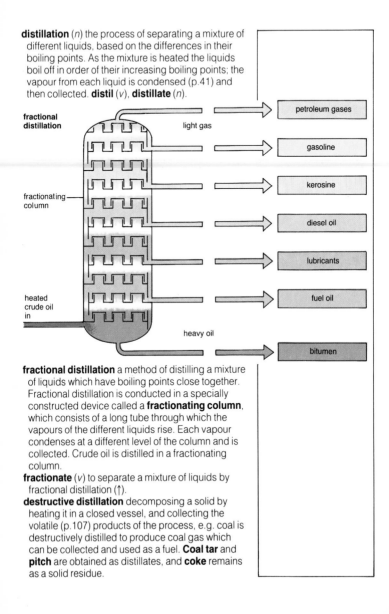

fractional distillation

light gas

fractionating column

heated crude oil in

heavy oil

petroleum gases

gasoline

kerosine

diesel oil

lubricants

fuel oil

bitumen

fractional distillation a method of distilling a mixture of liquids which have boiling points close together. Fractional distillation is conducted in a specially constructed device called a **fractionating column**, which consists of a long tube through which the vapours of the different liquids rise. Each vapour condenses at a different level of the column and is collected. Crude oil is distilled in a fractionating column.

fractionate (*v*) to separate a mixture of liquids by fractional distillation (↑).

destructive distillation decomposing a solid by heating it in a closed vessel, and collecting the volatile (p.107) products of the process, e.g. coal is destructively distilled to produce coal gas which can be collected and used as a fuel. **Coal tar** and **pitch** are obtained as distillates, and **coke** remains as a solid residue.

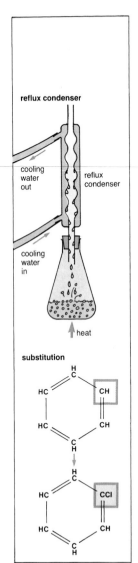

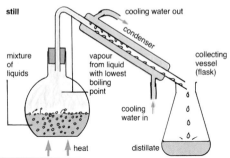

still (*n*) an apparatus used for the distillation (↑) of liquids, especially ethanol or petroleum.

reflux condenser a device in which the vapour produced by a boiling liquid is condensed (p.41) and falls back into that liquid. It can be used to raise the temperature of a chemical reaction without the liquid boiling dry. **reflux** (*v*).

cracking (*n*) the decomposition of organic (p.113) compounds, especially mineral oils, by heat. The purpose of cracking is to convert mineral oils of high boiling point into ones of lower boiling point more suitable for use in petrol engines. Petrol (gasoline) is produced by this method.

fermentation (*n*) the chemical change brought about in carbohydrates by the action of living organisms, such as yeasts and bacteria; the organisms produce enzymes which catalyze the decomposition. The carbohydrate is broken down into ethanol (ethyl alcohol) and carbon dioxide.

hydrolysis (*n*) the chemical decomposition (p.45) of an organic compound by water. The salts of weak acids or weak bases are partly hydrolyzed by water to form an alkaline or acidic solution. Esters (p.116) are hydrolyzed by water to form an alcohol and an acid. **hydrolyze** (*v*), **hydrolytic** (*adj*).

saponification (*n*) the hydrolysis (↑) of an ester heated with an alkali; it produces an alcohol and a salt of the organic acid or acids present.

substitution (*n*) the replacing (p.46) of an atom or group of atoms of a molecule by an atom or group of atoms of a different element, e.g. replacing one of the hydrogens of benzene by a chlorine atom to form chlorobenzene.

synthesis (*n*) the combining of elements or simple compounds to form a complex compound, e.g. the synthesis of dyes from aniline.

synthetic (*adj*) describes substances or materials that are produced by chemical processes, as distinct from those that occur naturally or are obtained from natural products. Plastic and nylon are synthetic materials.

nylon (*n*) a polymer (↓) consisting of amide groups (−CHNH−) recurring along a chain of carbon atoms. Nylon is a term used to describe all such polymers, despite the fact that they may have different physical properties. The most familiar form of nylon is that used for threads.

vulcanization (*n*) the process of heating natural rubber with sulphur to make it harder so that it keeps its shape under stress and is less elastic. Tyres are made from vulcanized rubber.

isomer (*n*) any of two or more compounds that have the same molecular formula, i.e. the same number of atoms of each element combined together, but have different physical or chemical properties due to differences in the structure of their molecules, e.g. butane and methyl propane are isomers.

alcohol (*n*) an organic compound derived from a hydrocarbon (p.113) but with one or more hydrogen atoms replaced by hydroxyl groups (p.110), e.g. ethanol is an alcohol; it is derived from ethane.

ester (*n*) an organic compound formed as a result of a reaction between an acid, usually an organic acid, and an alcohol. Water is a by-product of the reaction, e.g. methanol and methanoic acid form methyl methanoate. **esterification** (*n*).

soap (*n*) a metallic salt of a fatty acid. Soaps are produced by the action of sodium or potassium hydroxide on fats or oil.

detergent (*n*) a substance used for cleaning, removing dirt or oil, etc. Detergents have long chain hydrocarbon molecules with an acidic group at one end of the chain. The hydrocarbon chains are insoluble in water but soluble in grease and oil; the acidic group is soluble in water, but insoluble in grease and oil. The acidic group is usually a sulphonic acid. Soaps are included in detergents, although the term is most frequently used for synthetic sulphonates.

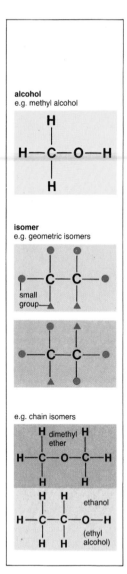

alcohol
e.g. methyl alcohol

isomer
e.g. geometric isomers

small group

e.g. chain isomers

dimethyl ether

ethanol
(ethyl alcohol)

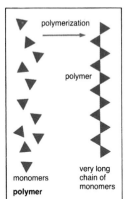

monomers

very long chain of monomers

polymer

polymer (*n*) a very long chain molecule formed by the chemical linking of smaller molecules of the same substance, called monomers, e.g. polyethylene (polythene), formula $(CH_2—CH_2)_n$ is formed by joining together many molecules of the monomer ethene (ethylene), formula $CH_2—CH_2$. **polymerize** (*v*), **polymeric** (*adj*).

polymerization (*n*) the process of forming a polymer by joining many simple molecules (monomers) into a long chain compound. **polymerized** (*adj*).

polythene (*n*) a tough, flexible plastic material made by the polymerization of ethene, C_2H_4. Previously known as polyethylene or alkathene. Used as an insulating material and for any purpose where a chemically-resistant material is needed.

polystyrene (*n*) a plastic made by the polymerization of styrene (phenylethene $C_6H_5.CH:CH_2$). It has good insulating properties.

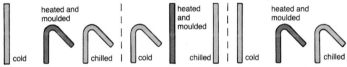

substance formed from polymers

thermoplastic

thermoplastic (*n*) a substance, formed of polymers, that is soft and easily moulded when subjected to heat, and hardens when chilled. It can be repeatedly remelted without changing its chemical properties, e.g. polythene and perspex are thermoplastics.

thermosetting (*adj*) describes a substance, formed of polymers, that is softened and easily moulded on first heating, but hardens as a chemical change occurs. The substance cannot be softened again by heat.

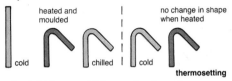

thermosetting

plastic² (*n*) a material that can be shaped or moulded by heat, but is stable and retains its shape at normal room temperature. Thermoplastics and thermosetting substances are plastic.

industry (*n*) any business or process (↓) concerned with the construction or manufacture (↓) of materials, products, substances or articles that are to be sold, e.g. the food industry; the electronics industry; the coal industry; the steel industry. **industrial** (*adj*).

manufacture (*v*) to make a chemical substance or a material in large quantities by a process which has a number of stages. A continuous supply of the substance or material is produced, e.g. to manufacture nitric acid by the Haber process. **manufacture** (*n*).

raw material a material still in its natural state, i.e. one that has not been processed or manufactured, e.g. bauxite is a raw material for making aluminium.

by-product (*n*) anything produced or obtained in the manufacturing process that is not the main, intended product, e.g. chlorine is a by-product from the manufacture of sodium hydroxide from sodium chloride.

waste product a product of a process for which no industrial use can be found.

process¹ (*n*) a set of consecutive events dependent upon one another and directed towards one ultimate objective. In chemistry, a process is considered to be the stages involved in the preparation, manufacture or isolation of a compound, e.g. the Solvay process for manufacturing sodium carbonate.

heat exchanger a part of a manufacturing process, or an energy generator, in which heat is transferred from one material to another, either to prevent loss of heat in the process, or to make use of the heat. In some processes the hot products of a reaction are used to heat up the incoming gases to conserve fuel, e.g. cold oxygen and sulphur dioxide in the contact process for the manufacture of sulphuric acid are heated by the outgoing product before entering the catalyst chamber.

fuel (*n*) any material that can be burned to provide heat or power. Natural gas, coal and oil are examples of fuel. Most fuels are carbon compounds.

fossil fuel a fuel formed in deposits in the ground from the remains of decayed plants and animals, e.g. coal, crude oil (petroleum).

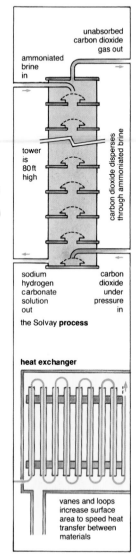

unabsorbed carbon dioxide gas out

ammoniated brine in

tower is 80 ft high

carbon dioxide disperses through ammoniated brine

sodium hydrogen carbonate solution out

carbon dioxide under pressure in

the Solvay **process**

heat exchanger

vanes and loops increase surface area to speed heat transfer between materials

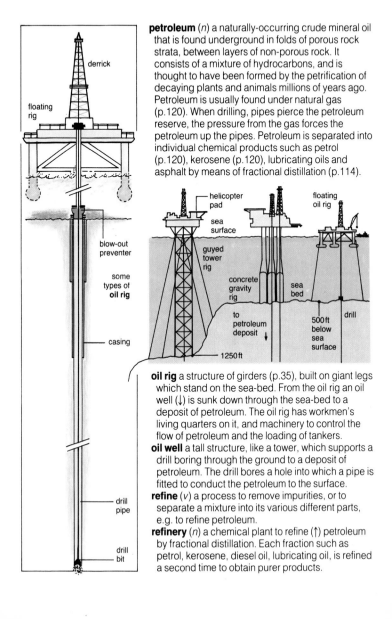

petroleum (*n*) a naturally-occurring crude mineral oil that is found underground in folds of porous rock strata, between layers of non-porous rock. It consists of a mixture of hydrocarbons, and is thought to have been formed by the petrification of decaying plants and animals millions of years ago. Petroleum is usually found under natural gas (p.120). When drilling, pipes pierce the petroleum reserve, the pressure from the gas forces the petroleum up the pipes. Petroleum is separated into individual chemical products such as petrol (p.120), kerosene (p.120), lubricating oils and asphalt by means of fractional distillation (p.114).

oil rig a structure of girders (p.35), built on giant legs which stand on the sea-bed. From the oil rig an oil well (↓) is sunk down through the sea-bed to a deposit of petroleum. The oil rig has workmen's living quarters on it, and machinery to control the flow of petroleum and the loading of tankers.

oil well a tall structure, like a tower, which supports a drill boring through the ground to a deposit of petroleum. The drill bores a hole into which a pipe is fitted to conduct the petroleum to the surface.

refine (*v*) a process to remove impurities, or to separate a mixture into its various different parts, e.g. to refine petroleum.

refinery (*n*) a chemical plant to refine (↑) petroleum by fractional distillation. Each fraction such as petrol, kerosene, diesel oil, lubricating oil, is refined a second time to obtain purer products.

natural gas a naturally-occurring gas that is formed in pockets above underground mineral oil deposits. Natural gas consists of hydrocarbon gases, mainly methane. It is used as a fuel (p.118).

petrol (*n*) a mixture of hydrocarbons (p.113) with a boiling point (p.43) between 20°C and 200°C, depending on the mixture. Petrol is used in internal combustion engines (p.18) and as a solvent for paints.

gasoline (*n*) alternative name for **petrol** (↑).

kerosene (*n*) a mixture of hydrocarbons obtained from petroleum; it has a boiling point between 175°C and 275°C. Kerosene is used as a fuel in jet engines and gas turbines, as well as in domestic cookers.

diesel oil a mixture of hydrocarbons obtained from petroleum; it has a boiling point between 250°C and 400°C. Diesel oil is used in diesel internal combustion engines in motor cars, lorries, train engines and ships' engines.

petrochemical (*n*) a chemical substance manufactured from raw materials (p.118) obtained from petroleum (p.119), e.g. benzene is obtained from hydrocarbons produced in a refinery (p.119); styrene (phenylethene) and aniline (phenylamine) are manufactured from benzene; both styrene and aniline are petrochemicals.

metallurgist (*n*) a person engaged in the scientific study and technology of metals, such as the extraction of metals from their ores, the production of alloys, studying the use of the properties of metals and alloys to meet specific requirements. **metallurgy** (*n*).

smelt (*v*) to obtain a metal from its ore by heating it with a suitable substance (usually coke) to a very high temperature, e.g. to smelt iron ore in a blast furnace (↓) to produce iron. **smelting** (*n*).

furnace (*n*) an enclosed chamber capable of producing very high temperatures by means of a fuel or an electric current. Furnaces are used for heating buildings or for smelting ores and refining metals.

blast furnace a type of furnace (↑) used primarily for separating iron from its ores. It produces the intense heat needed to smelt the iron by means of a blast of hot air forced into the furnace from below.

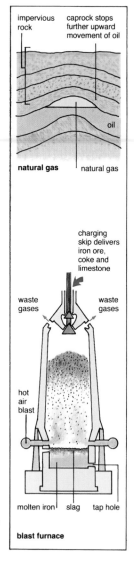

impervious rock caprock stops further upward movement of oil

oil

natural gas natural gas

charging skip delivers iron ore, coke and limestone

waste gases waste gases

hot air blast

molten iron slag tap hole

blast furnace

steel (*n*) iron containing 0.1% to 1.5% carbon. Other elements can be added to molten steel to produce alloy steels, e.g. stainless steel contains 10–20% chromium and 8–20% nickel. Steel and alloy steels are harder, more elastic, and can be given much more varied properties than iron.

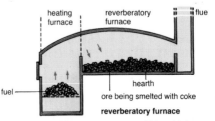

reverberatory furnace

reverberatory furnace a furnace (↑) in which an ore or metal is heated by flames and currents of hot air reflected downward from the roof of the furnace.

slag (*n*) a floating film of non-metallic waste that forms on the surface of the molten metal during the smelting of ores. It consists of calcium silicate, and other mineral material.

kiln (*n*) a furnace for making lime, bricks or pottery.

mould[1] (*n*) a hollow vessel into which a liquid or powdered plastic is poured; a solid part is lowered in, and this, together with the vessel, makes the required shape. The contents are usually heated and put under pressure. The result is a solid of a particular shape given to it by the mould. **mould** (*v*).

injection moulding making solid plastic (p.117) objects by forcing hot liquid plastic material into a cold mould (↑).

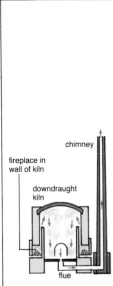

fireplace in wall of kiln

chimney

downdraught kiln

flue

heated gases follow path of →

kiln

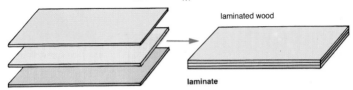

glue spread between wood layers and boards pressed together

laminated wood

laminate

laminate (*v*) to make solid sheets of material from thin layers of wood, paper or cardboard fixed together by layers of plastic. Laminated wood is stronger than plywood which has thin layers of wood glued together.

astronomy (*n*) the study of the stars and planets, with measurements of their motion, relative distances, chemical composition, physical conditions. **astronomical** (*adj*).

radio astronomy the study of radio waves emitted by bodies in space, such as radio stars and the Sun. A disc-shaped aerial is used to listen to the radio waves; it also determines the direction of the source.

observatory (*n*) a place where astronomers work, using telescopes to observe and to photograph the stars. By observing starlight, the temperature of a star, its chemical composition, and its distance can be determined. Radio telescopes are also used for observations.

stellar (*adj*) of a star, or concerning the stars, e.g. stellar evolution is a theory of the origin and development of stars.

solar (*adj*) concerning the Sun, or coming from the Sun, e.g. solar energy is energy radiated by the Sun.

lunar (*adj*) of, or concerning, the Moon, e.g. lunar landscape; lunar module.

sidereal (*adj*) of measurements made relative to the stars, e.g. a sidereal month is the time taken for the Moon to complete one revolution, relative to a given star, as observed from the Earth.

universe (*n*) everything that exists; the totality of all matter, energy and space. **universal** (*adj*).

galaxy (*n*) a massive collection of stars, dust and gases. There are millions of stars in each galaxy, and millions of galaxies in the universe (↑). Galaxies are separated from each other by vast distances. The Earth, Sun and the planets are in the Milky Way galaxy, which is spiral-shaped.

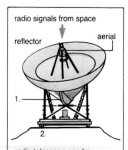

radio signals from space

reflector

aerial

1.

2.

radio telescope can be turned to any angle –
1. vertical angle control
2. horizontal angle control

radio astronomy

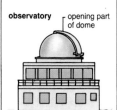

observatory opening part of dome

telescope is mounted below centre of dome, part of which slides open to give a clear view of the sky

galaxy
left – elliptical galaxy
right – barred spiral galaxy

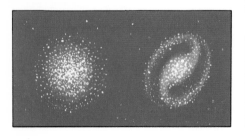

light year a measure of distance in the universe. One light year is the distance that a ray of light in a vacuum would travel in one year, i.e. approximately 9.5 million million km. The Milky Way galaxy (↑) is approximately 100 000 light years across.

sun (*n*) (1) a large, very hot, spherical star. (2) the Sun in our solar system. It is a source of light, heat and energy; Earth and the planets revolve around it. The Sun has a mass of approximately 2×10^{30} kg, a diameter of 1 392 000 km, and a surface temperature of about 6000°C.

eclipse (*n*) the dark shadow that is cast on one body in space when light travelling towards it is blocked by another body in space. An eclipse of the Sun (solar eclipse) occurs when the Moon passes between the Sun and the Earth and blocks the light reaching the Earth. An eclipse of the Moon (lunar eclipse) occurs when the Earth comes between the Sun and the Moon and blocks the light reaching the Moon.

partial/total eclipse

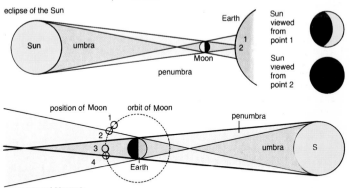

eclipse of the Sun

Sun

umbra

penumbra

Earth

Moon

1
2

Sun viewed from point 1

Sun viewed from point 2

position of Moon

orbit of Moon

penumbra

1
2
3
4

Earth

umbra

S

appearance of Moon at:

1 no shadow falls on Moon

2 partial eclipse

3 total eclipse

4 partial eclipse

partial eclipse occurs when one body in space, or a part of that body, is in the penumbra (p.55) of the shadow cast by another body in space. Observers in the penumbra will see the light source only partially obscured.

total eclipse occurs when one body in space, or a part of that body, is in the umbra (p.55) of the shadow cast by another body in space. Observers in the umbra will see the light source completely obscured.

corona (*n*) (1) a set of coloured rings seen around
the Sun or the Moon when the atmosphere is misty.
(2) an irregular white band around the Sun, seen
during a total eclipse; it is the Sun's atmosphere.
coronal (*adj*).

sunspot (*n*) a large patch, black in the middle with a
lighter part round it, that appears on the surface of
the Sun. A sunspot appears to move across the
surface because the Sun rotates. Sunspots appear
and disappear, but their number reaches a
maximum every eleven years; magnetic storms and
disruptions of radio communications are associated
with this maximum number of sunspots. A sunspot
is a cooler part of the Sun's surface.

solar energy the surface of the Sun is at a
temperature of 6000°C, and the interior is at a
temperature of 13 000 000°C. The internal
temperature is sufficiently high for thermonuclear
fusion to take place and hydrogen is transmuted to
helium with a release of vast quantities of energy.
This energy is released as radiation of
electromagnetic waves of all frequencies, including
radio-frequencies. Short outbursts of high
temperatures are observed as bright patches,
called **solar flares**, during which time protons and
electrons, and intense radio-frequency waves are
emitted. The solar energy emitted by the Sun is
equivalent to approximately 1400 J per square
metre on the Earth's surface, but most of this energy
is absorbed by the atmosphere.

solar cell a photovoltaic cell is used. When light
above a limiting frequency falls on a junction of two
dissimilar metals, or of a metal and a
semiconductor, or two dissimilar semiconductors,
an electromotive force is generated. A
metal/semiconductor cell is shown in the diagram.
Solar cells use such a junction or a p-n
semiconductor junction set in a silicon crystal.
Sunlight contains electromagnetic waves of
sufficiently high frequency to develop an e.m.f.
across the cell, caused by electrons diffusing
across the junction.

solar panel a battery of solar cells (↑) which provides
enough electrical power to operate electronic
equipment in artificial satellites and space probes
(p.133).

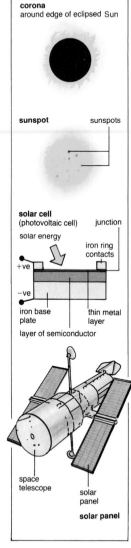

corona
around edge of eclipsed Sun

sunspot   sunspots

solar cell
(photovoltaic cell)   junction
solar energy
                               iron ring
                               contacts
+ve
−ve
iron base   thin metal
plate   layer
layer of semiconductor

space
telescope
                    solar
                    panel
                solar panel

cosmic rays radiation from outer space entering the Earth's atmosphere. The primary rays consist mainly of protons with some electrons and alpha (p.66) particles; all the particles have high energy. The primary rays on entering the Earth's atmosphere collide with gaseous atoms and produce secondary rays which, in turn, give rise to 'hard' and 'soft' radiation. Hard radiation includes gamma rays (p.66) and high-energy electrons; it has great penetrating power, penetrating deep into the Earth. High-energy electrons give rise to cosmic ray showers in which photons (p.134) and positrons (a positively charged particle with the same mass as an electron) are generated; penetrating showers can penetrate up to 20 cm of lead. Soft radiation consists mainly of electrons and has a low penetrating power. The origin of cosmic rays is uncertain, although some rays appear to be emitted by the Sun.

luminous (*adj*) describes something that gives out light, e.g. the Sun is a luminous body.

magnitude² (*n*) a measure of the brightness of a star as compared with a group of standard stars. It is expressed as a number on a scale. The brightest stars are of the first magnitude, and the faintest stars are of the sixth magnitude.

star (*n*) a luminous (↑) body, formed of very hot gases, that is found in space. It remains in a fixed position relative to other stars. The Sun is a star.

constellation
e.g. Ursa Major, showing stars of different magnitude

nebula

constellation (*n*) a number of fixed stars that form a pattern or shape in the sky, and are recognized and named as a group, e.g. the constellation of Gemini; the constellation of Leo.

nebula (*n. pl. nebulae*) a luminous (↑) cluster of dust and very hot gases found in space. Stars are formed from nebulae.

nova (*n. pl. novae*) one of a pair of double stars (p.125). It is bright for a few days, then faint again for a longer period. A nova has used up its hydrogen and is becoming smaller. It ejects a cloud of gas becoming 5000–10000 times brighter, and then becomes faint.

supernova (*n. pl. supernovae*) a star which has used up all its hydrogen and contracts. This produces a high temperature giving rise to thermonuclear reactions which produce heavy elements by nuclear fusion; these elements have relative atomic masses greater than 40. The star collapses inwards and explodes, becoming 10^8 times brighter than the Sun. The residue is a white dwarf (↓). The explosion of a supernova is a rare event.

white dwarf a small, highly dense star with a low luminosity; because of its size it has a high surface temperature and thus appears white. It is so dense that 1 cm^3 of the star has a mass of several tonnes. A white dwarf is a star near the end of its life.

red giant a large star with a reddish appearance and a low luminosity. It has consumed 80–90% of its hydrogen and the rate of consumption of hydrogen is increasing, being greater than the rate of main sequence stars. Eventually, a red giant contracts and becomes a white dwarf (↑) when all its hydrogen has been consumed.

pulsar (*n*) a star which emits radio-frequency radiation in brief pulses at regular intervals. A few pulsars also emit pulses of light. Pulsars are thought to be stars which have consumed all their nuclear fuel, have become highly compressed and consist of neutrons surrounded by a thin shell. The star then rotates emitting pulses of electromagnetic radiation; it is at the end of its evolutionary process, and its density is 10^7 times greater than that of a white dwarf (↑).

quasar (*n*) an extra-galactic source of high energy electromagnetic radiation. It emits powerful radio-frequency waves, and quasars were discovered from such radiation. Some quasars are visible, and from optical measurements are found to be travelling away from the Earth at very high velocities. A quasar is hundreds of times smaller than a galaxy (p.122), but emits radiation hundreds of times greater than that emitted by a galaxy.

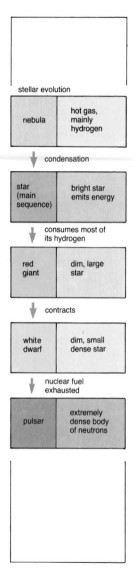

stellar evolution

| nebula | hot gas, mainly hydrogen |

↓ condensation

| star (main sequence) | bright star emits energy |

↓ consumes most of its hydrogen

| red giant | dim, large star |

↓ contracts

| white dwarf | dim, small dense star |

↓ nuclear fuel exhausted

| pulsar | extremely dense body of neutrons |

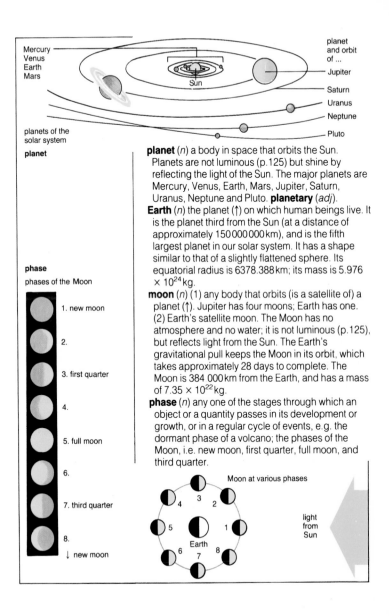

Mercury
Venus
Earth
Mars

planet
and orbit
of ...

Sun

Jupiter

Saturn

Uranus

Neptune

Pluto

planets of the
solar system

planet

phase

phases of the Moon

1. new moon

2.

3. first quarter

4.

5. full moon

6.

7. third quarter

8.

↓ new moon

planet (*n*) a body in space that orbits the Sun.
Planets are not luminous (p.125) but shine by
reflecting the light of the Sun. The major planets are
Mercury, Venus, Earth, Mars, Jupiter, Saturn,
Uranus, Neptune and Pluto. **planetary** (*adj*).

Earth (*n*) the planet (↑) on which human beings live. It
is the planet third from the Sun (at a distance of
approximately 150 000 000 km), and is the fifth
largest planet in our solar system. It has a shape
similar to that of a slightly flattened sphere. Its
equatorial radius is 6378.388 km; its mass is 5.976
$\times 10^{24}$ kg.

moon (*n*) (1) any body that orbits (is a satellite of) a
planet (↑). Jupiter has four moons; Earth has one.
(2) Earth's satellite moon. The Moon has no
atmosphere and no water; it is not luminous (p.125),
but reflects light from the Sun. The Earth's
gravitational pull keeps the Moon in its orbit, which
takes approximately 28 days to complete. The
Moon is 384 000 km from the Earth, and has a mass
of 7.35×10^{22} kg.

phase (*n*) any one of the stages through which an
object or a quantity passes in its development or
growth, or in a regular cycle of events, e.g. the
dormant phase of a volcano; the phases of the
Moon, i.e. new moon, first quarter, full moon, and
third quarter.

Moon at various phases

Earth

light
from
Sun

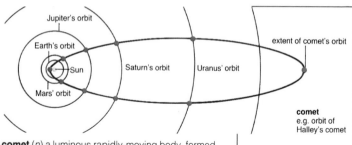

Jupiter's orbit

Earth's orbit

extent of comet's orbit

Sun

Saturn's orbit   Uranus' orbit

Mars' orbit

comet
e.g. orbit of
Halley's comet

comet (*n*) a luminous rapidly-moving body, formed of dust and hot gases, that orbits the Sun. Comets have a nucleus similar to that of a star (p.125), and a bright, streaming tail that always points away from the Sun. Their mass is very small, though they can reach a very great size.

meteor (*n*) a small solid body that is heated by friction when it enters the Earth's atmosphere, and as a result is burnt up. Small meteors are burnt up completely to gases; large meteors do not burn up completely and fall to Earth as **meteorites**.

black hole an area in outer space that attracts everything near it, including light, and pulls it into the black hole. A black hole is thought to be the residue of a collapsed, large star with a density so high it exerts a gravitational force on light and other electromagnetic radiations.

asteroid (*n*) a small planet in orbit around the Sun between the orbits of Mars and Jupiter. There are about 1500 asteroids forming a belt around the Sun. All asteroids have a diameter less than 500 km.

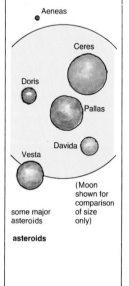

Aeneas

Ceres

Doris

Pallas

Davida

Vesta

(Moon
shown
for
comparison
of size
only)

some major
asteroids

asteroids

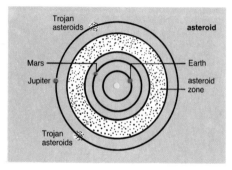

Trojan
asteroids

asteroid

Mars

Earth

Jupiter

asteroid
zone

Trojan
asteroids

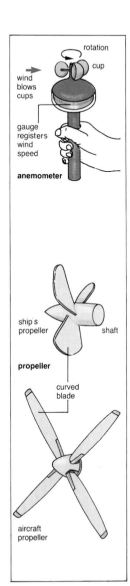

rotation

cup

wind blows cups

gauge registers wind speed

anemometer

ship's propeller

shaft

propeller

curved blade

aircraft propeller

aeronautics (*n*) the study of the science of flight and the operation of aeroplanes and airships. Compare rocketry, the science of using rockets for operations in space.

navigation (*n*) the science of determining the position of a ship, an aircraft, or a rocket, from astronomical measurements, calculations based on geometry and time, and from using other aids such as maps, depth soundings, lighthouses, radio beacons and radar (p.103). Navigation also includes the guiding of the ship, aircraft or rocket in the correct direction at a suitable speed to reach a required destination. **navigate** (*v*), **navigational** (*adj*).

meteorology (*n*) the science of the Earth's atmospheric conditions, e.g. the study of the pressure (p.27), temperature (p.37), humidity (p.40) of the atmosphere in all areas of the Earth, and at all levels of the atmosphere. This study, together with measurements of wind strength, observation of cloud formation, and the recording of changes in atmospheric measurements, allows the weather to be forecast. The forecasts are useful for navigation (↑) by ships and aircraft.

anemometer (*n*) an instrument for measuring the speed of the wind; it consists of three or four cups mounted on an axle. The wind rotates the cups and axle, and a gauge measures the speed or power of the wind.

aerodynamics (*n*) the study of the forces exerted by gases in motion or of the forces exerted by gases on bodies moving through the gases. In particular, the study of the forces acting on solid bodies, and the control of these bodies when in motion through air, e.g. aeroplanes. **aerodynamic** (*adj*).

propulsion (*n*) the action of maintaining an object in forward motion with the force causing the motion always to be present on the object, e.g. an aeroplane engine and its propeller (↓) provide propulsion for an aircraft. **propel** (*v*).

propeller (*n*) curved blades fitted to a rotating shaft. The blades are a section of a screw, so that the propeller bites into the water, or air, in the same way that a screw bites into wood. This propels the boat or the aircraft relative to the water or air. *See* **propulsion** (↑).

fuselage (*n*) the main body of an aeroplane; the wings, rudder and elevators are attached to it.

aerofoil (*n*) any surface, curved or flat, which makes use of aerodynamic forces, e.g. the wing of an aeroplane, curved on top and flat on the bottom, creates an upward thrust as it goes through the air. Wings, elevators, and the rudder are aerofoils on an aeroplane, i.e. all parts other than the fuselage (↑).

aileron (*n*) a movable edge section at the end of each wing of an aeroplane; when operated, ailerons make one wing rise and the other fall, causing the aeroplane to **bank**. Ailerons are used when changing direction.

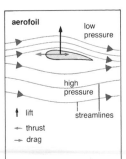

flap (*n*) a narrow, hinged section at the back of the wing of an aeroplane. When raised, it alters the curved shape of the wing, thus changing the upward thrust on the aerofoil (↑). Flaps are used when landing and taking off.

delta wing a type of aeroplane with a wing in the shape of the Greek letter delta (Δ).

flight (*n*) the action of flying. For an aeroplane, aerofoils (↑) (wings) are used to supply an upward thrust. Height above the ground is obtained from the use of elevators, and direction is altered by the use of ailerons and a rudder. Air turbulence causes the aeroplane to pitch, roll and yaw (↓). The aerofoils (wings, elevator, ailerons, rudder) correct these faults and guide the aeroplane to its destination at a suitable height.

drag (*n*) an aeroplane experiences a frictional force as it flies through the air. This force is drag. Drag and upthrust on a wing produce a resultant force.

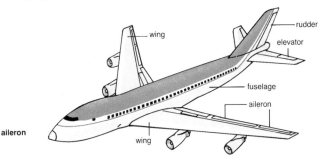

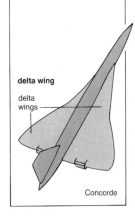

Concorde

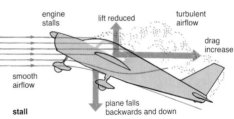

engine stalls

lift reduced

turbulent airflow

drag increase

smooth airflow

stall

plane falls backwards and down

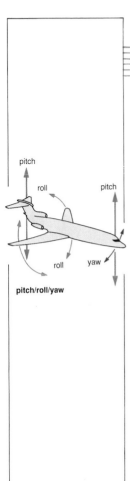

pitch

roll

pitch

pitch

roll

yaw

pitch/roll/yaw

stall (v) (1) to fly at too low a speed so that the upthrust on the wings is insufficient to support the aeroplane's weight; the aeroplane tends to fall backwards. (2) of an engine, to stop because the engine's power is too low to overcome the resistance to motion.

pitch² (v) to move up and down in an endwise motion, so the nose and rudder rise and fall. Pitching is corrected by the use of elevators.

roll (v) to turn about a horizontal axis, from side to side, so the wings of an aeroplane rise and fall. Rolling is corrected by the use of ailerons (↑).

yaw (v) to swing from side to side, about a vertical axis, away from the correct course. Yawing is corrected by use of the rudder.

supersonic (adj) describes a speed faster than that of sound under the same conditions, e.g. supersonic flight of an aircraft. Contrast **ultrasonic** (p.52).

Mach number the ratio of the speed of an object, or the speed of flow of a gas, to the speed of sound under the same conditions. Mach 1 indicates an object travelling at the speed of sound. If the Mach number of an object is greater than 1, the speed is supersonic (↑).

rocket (n) a device that is propelled by hot gases. The gases are forced out of a rear vent and drive the rocket forward or upwards. The rocket does not require air for combustion and so is not dependent upon the Earth's atmosphere and can travel in space.

propellant (n) refers to a substance that burns or explodes slowly and in a controlled manner in order to apply force to a projectile (p.132), e.g. gunpowder in a firework or flare; the solid or liquid fuel propellant in a rocket (↑).

projectile (*n*) a body which is driven by an impulse in a definite direction up into the air, e.g. a shell fired from a gun is a projectile. The impulse is usually an explosion. A projectile remains in the Earth's gravitational field, so that it eventually returns to Earth. **project** (*v*).

missile (*n*) a projectile (↑) used for scientific experiment, or for war. A ballistic missile is projected by an impulse, so its path depends on its initial direction and the strength of the impulse. A guided missile is projected by an impulse and has some form of mechanism which allows it to have its path altered during flight.

launch (*v*) to send a projectile, probe (↓), or spacecraft (↓) up into the atmosphere. Launching is carried out on a launch pad.

escape velocity the speed with which a satellite (↓) must be launched in order to escape from the retarding effect of the Earth's gravity; it is approximately 12km s^{-1}. Satellites are usually launched eastward, i.e. in the direction of rotation of the Earth, so that the satellite has the benefit of the rotational speed of the Earth.

orbit (1) (*n*) the path followed by a satellite or planet moving around another body in space under the effect of gravity, e.g. the orbit of the Moon round the Earth. (2) (*v*) to move in an orbit, e.g. the Moon orbits the Earth.

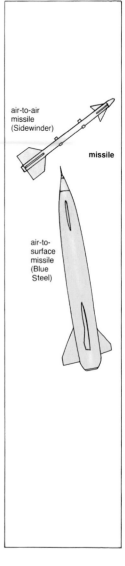

air-to-air missile (Sidewinder)

missile

air-to-surface missile (Blue Steel)

orbit

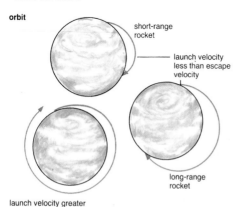

short-range rocket

launch velocity less than escape velocity

long-range rocket

launch velocity greater than escape velocity

orbital velocity

orbit

Earth

radius of orbit

orbital velocity v

satellite

orbital velocity the velocity of a satellite in orbit. The Earth's gravitational field extends out into space and acts on a satellite, attracting it to Earth. The velocity of the satellite gives it a centrifugal force, acting against gravity. In orbit, these two forces acting on the satellite are equal. If v is the orbital velocity, r the orbit radius from the centre of the Earth, and T the period of revolution, then $v \propto 1/\sqrt{r}$ and $T \propto \sqrt{r^3}$. The orbital velocity determines the height of the satellite and its period of revolution; the greater the height, the lower the orbital velocity, and the longer the period of revolution.

geostationary orbit an orbit in which a satellite revolves around the Earth so that it is always above the same terrestrial position; it does this by having a period of revolution of 1 day.

satellite (*n*) (1) a body in space which orbits a larger body, and is held in orbit by gravitational attraction, e.g. the Earth is a satellite of the Sun. (2) a man-made object put into orbit around a large body in space, such as the Moon or the Earth.

probe (*n*) a rocket carrying many different kinds of measuring instruments, and often cameras to take photographs, used to examine the conditions on a body in space. A lunar probe examines the Moon, planetary probes observe planets. A probe is powered by solar panels (p.124) or by nuclear energy, and is controlled by telecommunications from Earth. The probe is unmanned.

spacecraft (*n*) a means of transport for astronauts going into space; it is powered by rockets. It is divided into modules.

module (*n*) a section that can be removed from a device, a machine, or a structure, e.g. a spacecraft is divided into modules. A module has a function to perform which is part of the function of the whole structure, e.g. a spacecraft can contain a command module, a service module and a landing module. Only the command module returns to Earth with the crew.

astronaut (*n*) a member of the crew of a spacecraft. Astronauts control the motion of the spacecraft, take observations and make measurements in space.

cosmonaut (*n*) the Russian term for an astronaut (↑).

spacecraft shuttle

photon (*n*) a photon can be regarded as a particle associated with an electromagnetic wave; it is the smallest packet of energy supplied by an e.m. radiation. If ν is the frequency of the radiation, and *h* is Planck's constant, then ν*h* is the energy of a photon. A photon has the speed of light (3 × $10^8 ms^{-1}$ in a vacuum), no electrical charge, and zero mass at rest. A more restricted meaning of photon is a particle of light.

laser
amplification of
stimulated emission

● photon
● excited neon atom
○ neon atom

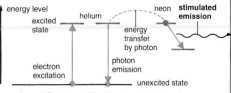

laser (*acro*) **l**ight **a**mplification by **s**timulated **e**mission of **r**adiation. If an atom absorbs energy an electron is raised to a higher energy level in an outer shell; when the electron falls back to its original level it emits a photon (↑) of a particular frequency, equivalent to the energy difference. For some energy levels, a photon, of the same frequency as the emitted photon, on colliding with an

waves reflected backwards and forwards

laser beam

totally silvered end
glass tube
power source
99% silvered end
helium and neon gases

excited atom (i.e. with an electron in the higher energy level) causes two photons to be emitted. This is **stimulated emission**. In a helium-neon laser, helium atoms are excited by electrons in a low-pressure glass discharge tube (p.98). The helium atoms emit photons which excite neon atoms, because both atoms have the same energy level. Photons collide with the excited neon atoms and stimulated emission takes place. One end of the glass tube containing the gases helium and neon is totally silvered, the other end is 99% silvered; both ends are optically flat and exactly parallel. The electromagnetic waves, produced by stimulated emission, are reflected backwards and

all waves same frequency

waves in phase
(in coherent light)

waves out of phase
(in incoherent light)

forwards between the silvered ends, becoming
more intense as more photon collisions take place;
the light has been 'amplified'. When the e.m. waves
are sufficiently intense, they pass through the
partially silvered end forming a continuous, intense
beam of light. The light is monochromatic, i.e. all the
waves are of the same wavelength, and it is also
coherent. Coherent light has every wave in phase;
light emitted from sources other than lasers,
e.g. sunlight, is incoherent. Coherent light is
needed to form interference patterns.

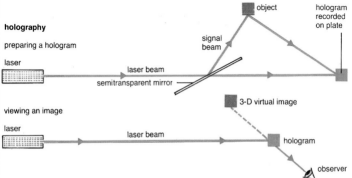

holography (*n*) the production of three-dimensional
visible images by the use of a laser (↑). A laser
beam is divided in two by a semitransparent mirror
to form a signal beam and a reference beam. The
signal beam is diffracted by an object and falls on a
photographic plate. The reference beam falls
directly onto the photographic plate. The two
beams produce interference patterns which are
recorded on the photographic plate. This record is
a **hologram**; no picture is visible, and the record is
a pattern of fine lines of the interference of the two
beams. The image is reproduced by illuminating the
hologram with the same laser light, or coherent light
of the same frequency, and looking through the
hologram as if looking through a window. A virtual
image is observed and it is three-dimensional.
hologram (*n*) a photographic plate on which a
pattern is printed containing the information for
reproducing an image by holography (↑).

robot (*n*) a mechanical device that is controlled by a computer to carry out simple tasks consisting of repeated operations, e.g. drilling holes in steel plates on an assembly line. A robot is equipped with sensors (↓) to detect the objects on which it operates. The sensors cause the computer to react and the computer program guides the robot in its task. **robotics** (*n*).

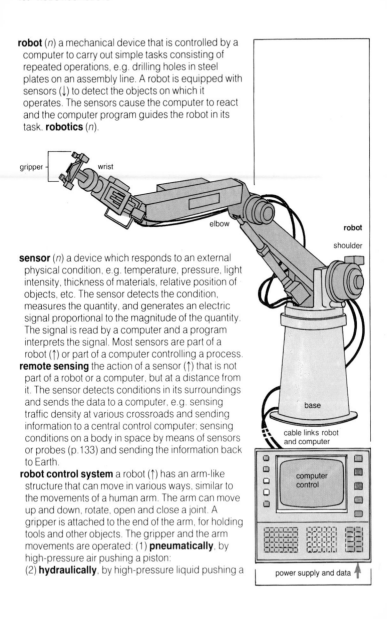

gripper | wrist

elbow

robot

shoulder

sensor (*n*) a device which responds to an external physical condition, e.g. temperature, pressure, light intensity, thickness of materials, relative position of objects, etc. The sensor detects the condition, measures the quantity, and generates an electric signal proportional to the magnitude of the quantity. The signal is read by a computer and a program interprets the signal. Most sensors are part of a robot (↑) or part of a computer controlling a process.

remote sensing the action of a sensor (↑) that is not part of a robot or a computer, but at a distance from it. The sensor detects conditions in its surroundings and sends the data to a computer, e.g. sensing traffic density at various crossroads and sending information to a central control computer; sensing conditions on a body in space by means of sensors or probes (p.133) and sending the information back to Earth.

base

cable links robot and computer

robot control system a robot (↑) has an arm-like structure that can move in various ways, similar to the movements of a human arm. The arm can move up and down, rotate, open and close a joint. A gripper is attached to the end of the arm, for holding tools and other objects. The gripper and the arm movements are operated: (1) **pneumatically**, by high-pressure air pushing a piston: (2) **hydraulically**, by high-pressure liquid pushing a

computer control

power supply and data ↑

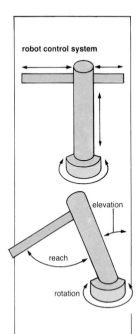

robot control system

elevation

reach

rotation

piston; or (3) **electrically**, by electric motors and relays (p.78). The exactness of position achieved by a robot on an object is called the robot's **repeatability**; many robots have a repeatability of a fraction of a millimetre. The motion of a robot is either **point-to-point**, or **continuous path**. Point-to-point motion needs a simple program for going from one position to another; continuous path motion needs a more complex program. The robot control system is a group of computers located in a cabinet away from the robot. Each type of movement is controlled by a small computer, and all the small computers are under the control of one large computer.

automation (*n*) the operation of a process using machines which control themselves, so that no operator is needed. Individual components are manufactured by robots (↑) using machine tools. Each component is moved from place to place by conveyor belts or by **A**utomatic **G**uided **V**ehicles (AGVs). Each robot and each AGV has its own robot control system (↑) programmed to carry out its own tasks; a central computer is in overall charge of the individual control systems.

robot assembly the building of a piece of machinery using robots to assemble and fit together all the various components. For such tasks, robots need sensors to estimate their grip on an object, so that they hold it not so hard that they damage the object, nor so gently that they risk dropping it. Electronic sensors are also needed for a visual identification system for different components. Robot assembly is not yet well developed.

artificial intelligence the capability of a robot (↑) or a computer when it can learn by correcting its mistakes and can adapt to situations not previously encountered. The emphasis is on the robot's ability to improve its performance on a process by learning from repeated experience.

remote control control exercised at a distance from the robot, computer or machine which is controlled, e.g. the remote control of an aeroplane from the ground, using radio signals, and electrical or electronic equipment; the supply of data from a remote terminal to a control computer, and the return of control signals from the main computer.

classify (*v*) to arrange objects or organisms into specified groups based upon their common characteristics (↓) or properties (p.25), e.g. the classification of an animal as a bird will depend upon it being warm-blooded, and possessing feathers, two wings, two legs and a beak. **classification** (*n*).

characteristic (*n*) a distinctive feature, shape or behaviour of an organism or substance, e.g. blue eyes, body hair. Combinations of characteristics may be used to classify organisms or objects, e.g. cold-blooded, smooth moist skin and having four limbs with five digits are characteristics of amphibians. **characteristic** (*adj*).

biological (*adj*) to do with biology or living things.

physiological (*adj*) to do with the function of an organism or part of an organism as opposed to its structure.

nature (*n*) (1) refers to all living and non-living objects (except man-made objects), and forms of energy contained in a system, e.g. animals, plants, mountains and oceans. (2) all the characteristics (↑) and properties of something that make it what it is, e.g. the nature of light, the nature of a bird, the nature of a metal.

kingdom (*n*) the first main division for the classification of living things, i.e. into the plant kingdom and the animal kingdom.

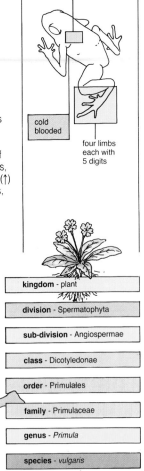

characteristic
e.g. characteristics of an amphibian

smooth moist skin

cold blooded

four limbs each with 5 digits

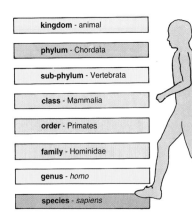

kingdom - animal	kingdom - plant
phylum - Chordata	division - Spermatophyta
sub-phylum - Vertebrata	sub-division - Angiospermae
class - Mammalia	class - Dicotyledonae
order - Primates	order - Primulales
family - Hominidae	family - Primulaceae
genus - *homo*	genus - *Primula*
species - *sapiens*	species - *vulgaris*

phylum (*n. pl. phyla*) a kingdom (↑) is divided into a number of phyla, e.g. the phylum of Chordata, animals with backbones. Plant phyla are usually called divisions.

genus (*n. pl. genera*) a group in the classification of organisms which consists of related species (↓). Members of the same genus may interbreed, the offspring is sterile in animals, but not always sterile in plants.

species (*n*) the smallest unit or group into which organisms are usually classified. Members of a species can breed amongst themselves to produce offspring who can in turn become parents. Individuals from different species cannot normally mate to produce young. If they do, the offspring are sterile and cannot reproduce, e.g. a horse and a donkey can breed to produce a mule but mules cannot reproduce.

animal (*n*) a living being which respires, excretes, grows and reproduces. Unlike plants, it moves from place to place and cannot make its own food from simple molecules, but obtains it by eating plants or other animals.

plant (*n*) a living organism which respires, grows and reproduces. Unlike animals plants cannot move from place to place; they make their own food from simple inorganic substances such as salts in solution and gases from the atmosphere.

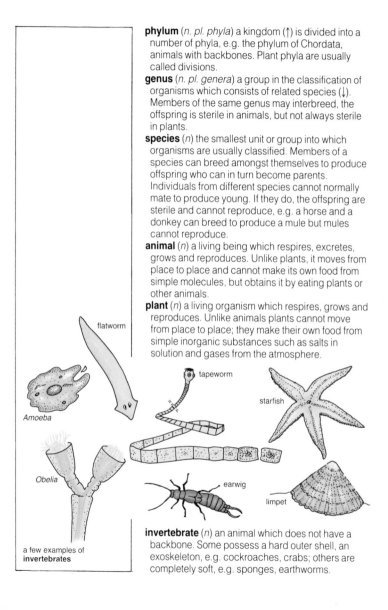

flatworm

tapeworm

starfish

Amoeba

Obelia

earwig

limpet

a few examples of **invertebrates**

invertebrate (*n*) an animal which does not have a backbone. Some possess a hard outer shell, an exoskeleton, e.g. cockroaches, crabs; others are completely soft, e.g. sponges, earthworms.

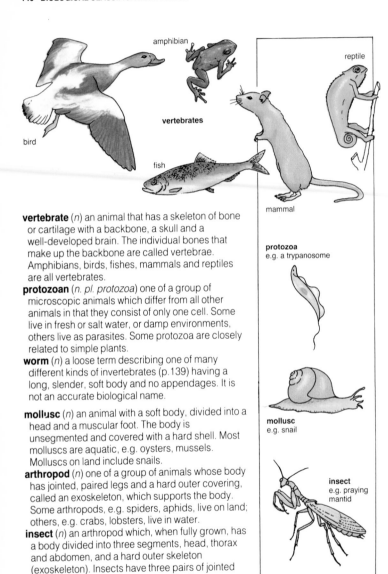

amphibian

reptile

vertebrates

bird

fish

mammal

protozoa
e.g. a trypanosome

mollusc
e.g. snail

insect
e.g. praying
mantid

vertebrate (*n*) an animal that has a skeleton of bone
or cartilage with a backbone, a skull and a
well-developed brain. The individual bones that
make up the backbone are called vertebrae.
Amphibians, birds, fishes, mammals and reptiles
are all vertebrates.

protozoan (*n. pl. protozoa*) one of a group of
microscopic animals which differ from all other
animals in that they consist of only one cell. Some
live in fresh or salt water, or damp environments,
others live as parasites. Some protozoa are closely
related to simple plants.

worm (*n*) a loose term describing one of many
different kinds of invertebrates (p.139) having a
long, slender, soft body and no appendages. It is
not an accurate biological name.

mollusc (*n*) an animal with a soft body, divided into a
head and a muscular foot. The body is
unsegmented and covered with a hard shell. Most
molluscs are aquatic, e.g. oysters, mussels.
Molluscs on land include snails.

arthropod (*n*) one of a group of animals whose body
has jointed, paired legs and a hard outer covering,
called an exoskeleton, which supports the body.
Some arthropods, e.g. spiders, aphids, live on land;
others, e.g. crabs, lobsters, live in water.

insect (*n*) an arthropod which, when fully grown, has
a body divided into three segments, head, thorax
and abdomen, and a hard outer skeleton
(exoskeleton). Insects have three pairs of jointed
legs attached to the thorax, one pair of antennae,

arachnid
e.g. scorpion

fish
e.g. goldfish

reptile
e.g. snake

and usually two pairs of wings (but not all have wings), e.g. ants, bees, beetles, butterflies, locusts, mosquitoes. During its life cycle it goes through changes in its body structure and way of life. *See* **metamorphosis** (p.146).

arachnid (*n*) an animal belonging to a sub-group of arthropods, mainly found on land. An arachnid has two body parts, head and thorax in one part, abdomen in the other. It has four pairs of jointed legs, simple eyes and no antennae, e.g. scorpion.

crustacean (*n*) an animal belonging to a sub-group of arthropods; most crustaceans are aquatic. A crustacean has a body divided into three parts, head, thorax, abdomen, with a hard exoskeleton (p.184) of lime. It has many pairs of jointed legs, and two pairs of antennae on the head, e.g. crab.

fish (*n.pl.*) a group of aquatic vertebrates (↑). The body is covered in scales, has fins and a tail. Fish breathe dissolved oxygen in water through gills; they have a heart with two chambers. Fish swim by means of their tail and fins.

amphibian (*n*) one of a group of cold-blooded vertebrate animals which live partly on land and partly in water. Amphibians have smooth, moist skin and four limbs each with five digits, e.g. frogs, toads. They lay their eggs, which are not protected by a shell, in water.

reptile (*n*) one of a group of cold-blooded vertebrate animals which live mainly on land, e.g. alligator, crocodile, snake. Reptiles have a dry, scaly skin and, except snakes, have four legs each with five digits. They lay eggs, covered with a tough shell membrane, on land, although some snakes give birth to live young.

bird (*n*) one of a group of warm-blooded vertebrate animals. Birds have a skin covered in feathers, beaks without teeth and scaly legs. They lay eggs protected by shells, e.g. eagles, geese, sparrows.

mammal (*n*) one of a group of warm-blooded vertebrate animals. Mammals have a skin covered in hair, a well-developed brain and a heart with four chambers. The young are born live, and feed on milk produced by the female's mammary glands. Some mammals, e.g. whales, dolphins and seals, live in the sea, but most live on land, e.g. humans, lions, elephants.

algae (*n*) a group of simple plants with a thallus instead of roots, leaves and stems; the thallus, a simple vegetative body, is either unicellular, or multicellular with chains of cells. All algae are aquatic, reproduce by binary division, and possess chlorophyll, e.g. spirogyra, seaweeds.

fungus (*n. pl. fungi*) a simple plant that contains no chlorophyll (p.223). It may consist of either a single cell or a mass of cellular, tube-like filaments called hyphae (p.144). Fungi usually feed off dead or decaying organisms and reproduce by means of spores produced in sporangia (p.144). **fungal** (*adj*).

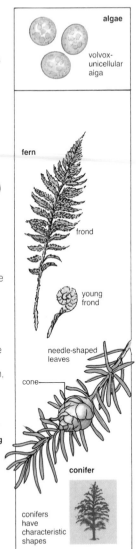

algae

volvox-
unicellular
aiga

fern

frond

young
frond

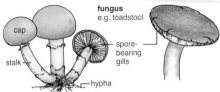

fungus
e.g. toadstool

cap

stalk

hypha

spore-
bearing
gills

fern (*n*) a spore-bearing plant with an underground perennial rhizome (p.225), conspicuous fronds (leaves) and vascular bundles (p.221) in stems. The spores are borne on the under-surface of the fronds.

conifer (*n*) a seed-bearing plant which is a tall, evergreen tree, found in temperate climates. A conifer does not produce flowers. Most conifers produce male and female cones. Naked seeds, covered with a thin membrane, are borne inside the leaf-like structures on the cone.

flowering plant a seed-bearing plant with root, stem, leaves and flowers. The seeds are enclosed in an ovary on the plant, the ovary becomes a fruit or a nut. Flowering plants are divided into two groups, monocotyledons (p.230) and dicotyledons (p.230).

needle-shaped
leaves

cone

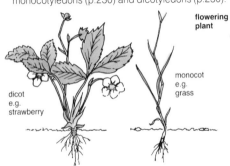

**flowering
plant**

dicot
e.g.
strawberry

monocot
e.g.
grass

conifer

conifers
have
characteristic
shapes

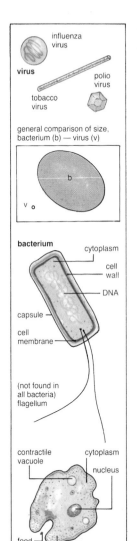

bacterium

cytoplasm
cell wall
DNA
capsule
cell membrane
(not found in all bacteria) flagellum

contractile vacuole
cytoplasm
nucleus
food
pseudopodium
amoeba

organism (*n*) any living object, including plants and animals, e.g. trees, fungi, dogs, bacteria.

virus (*n*) a parasite which causes disease by reproducing in living cells but is not normally considered to be living as some viruses have been crystallized. The diameter of a virus is between 0.03 μm, and 0.5 μm, much smaller than a bacterium (↓). Unlike bacteria, a virus cannot reproduce outside a living cell. **viral** (*adj*).

bacteriophage (*n*) a type of virus (↑) which attacks bacteria (↓).

bacterium (*n. pl. bacteria*) an organism (↑) consisting of a single cell (p.149) with no separate nucleus (p.150). Bacteria are neither plants nor animals and come in a variety of shapes from rod-like to spiral. They are between 0.5 μm and 2.0 μm in size. Some bacteria have flagella (p.153) for locomotion; spherical bacteria are unable to move unaided as they lack flagella. Bacteria reproduce asexually by binary fission (p.152), they are found in very large numbers everywhere. They are concerned with the decay of plant and animal tissues (p.150), e.g. sewage disposal; chemical changes of inorganic salts, such as nitrites to nitrates; manufacture of certain foods, such as cheese; the spread of disease, such as tuberculosis. **bacterial** (*adj*).

amoeba (*n*) a microscopic organism, with only one cell, containing cytoplasm, a nucleus and vacuoles (p.151). Its jelly-like body continually changes shape by movement of the cytoplasm. It feeds by extending pseudopodia (p.153) around a particle of food which is then absorbed inside a food vacuole. Amoeba reproduce asexually by binary fission. **amoebic** (*adj*), **amoeboid** (*adj*).

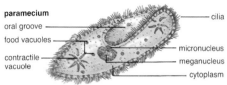

paramecium
cilia
oral groove
food vacuoles
micronucleus
contractile vacuole
meganucleus
cytoplasm

paramecium (*n*) a single-celled organism, shoe-like in shape. Its surface is covered with cilia (p.153) which propel the paramecium through the water by wave-like beating.

mould² (*n*) a green, grey or white growth of a fungus (p.142) on the surface of a living or dead organism, e.g. mould on cheese. It causes the decay of the organic matter and is found particularly on damp or decaying (↓) organisms. **mouldy** (*adj*).

hypha (*n. pl. hyphae*) the tube-like filament of a fungus (p.142). It is filled with cytoplasm (p.150) containing nuclei, but not separated by membranes or cell walls into cells. It grows at the tip and branches to form new hyphae. **hyphal** (*adj*).

mycelium (*n. pl. mycelia*) the tangled mass of hyphae (↑) which makes up the whole of a fungus other than the reproductive parts.

sporangium (*n. pl. sporangia*) the structure in a fungus that produces spores. It is carried on the end of a stalk growing up from the mycelium (↑).

spore (*n*) a reproductive cell produced in large numbers by some plants, fungi (p.142), bacteria (p.143), and protozoa (p.140). In plants the spores are formed by vegetative reproduction. The spores differ from seeds in not possessing an embryo, and under suitable conditions grow into a new individual.

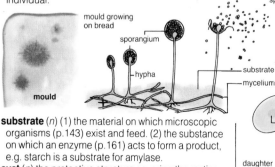

mould growing on bread

sporangium

spores

hypha

substrate

mycelium

mould

cell wall

cytoplasm

nucleus

daughter cell forms

bud forms

new cells form

yeast

substrate (*n*) (1) the material on which microscopic organisms (p.143) exist and feed. (2) the substance on which an enzyme (p.161) acts to form a product, e.g. starch is a substrate for amylase.

cyst (*n*) the protective structure covering the resting stage of an organism, formed when conditions are unfavourable. **cystic** (*adj*).

yeast (*n*) a single-celled fungus (p.142). It produces enzymes (p.161) which decompose sugars and starch forming alcohol and carbon dioxide (fermentation). Yeasts multiply asexually by budding. Different kinds of yeast act on different substrates (↑) and are used in the brewing, wine-making and baking industries.

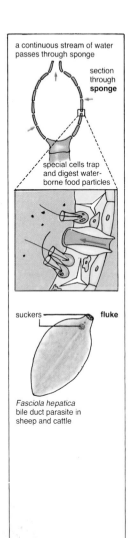

a continuous stream of water passes through sponge

section through **sponge**

special cells trap and digest water-borne food particles

suckers   **fluke**

Fasciola hepatica bile duct parasite in sheep and cattle

penicillium (*n*) a type of mould. The powerful antibiotic, penicillin, is made from it.

decay (*n*) the chemical processes which occur in plant and animal materials after death, due to the action of bacteria (p.143) and fungi (p.142). Decay decomposes the tissues and chemical compounds releasing carbon dioxide, ammonia and water.

sponge (*n*) a simple multicellular aquatic animal with a hollow body. It has tissues but no organs or nerves. Sponges are unable to move and usually live in colonies attached to rocks. They feed by using flagella (p.153) to draw in water containing food particles through small holes, and then expel the water through larger holes.

fluke (*n*) a small, flat, worm-like animal up to 1 cm long. Flukes are parasitic (p.240) and can cause serious diseases, e.g. bilharziasis in humans. They have a complicated life cycle with different structures in each stage.

vermiform (*adj*) shaped like a worm, e.g. vermiform appendix.

life history the changes that an organism goes through from the fertilized egg until death. It also refers to the changes that occur in a single stage (p.146) of a life cycle (↓), e.g. the life history of a tadpole. The individual life histories together make up the life cycle of an organism.

life cycle all the stages and changes that a living organism goes through from the fertilized egg until death, for example, the life cycle of a frog is egg–tadpole–adult frog–death; the life cycle of a housefly is egg–larva–pupa–imago–death.

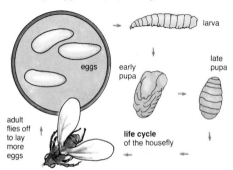

larva

eggs

early pupa

late pupa

adult flies off to lay more eggs

life cycle of the housefly

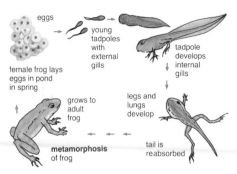

metamorphosis (*n*) the change of form that some animals undergo during their life cycle, e.g. butterflies and moths change from egg to larva (caterpillar) to pupa (chrysalis) to imago (adult). **Incomplete metamorphosis** in insects involves only nymph and adult stages, e.g. grasshopper; **complete metamorphosis** involves the change from larva to pupa to imago, e.g. butterfly, housefly, moth.

stage (*n*) a period, division or part of a process or life of an organism or object, e.g. the three stages in the life of a frog: egg–tadpole–adult frog.

butterfly (*n*) an insect (p.140) of the order Lepidoptera. The thorax (p.154) is divided into three segments, with a pair of jointed legs on each one and two pairs of large wings attached to the second and third segments. The wings and body are covered in small scales and are often brightly coloured. It undergoes complete metamorphosis (↑) during its life cycle. Butterflies feed on nectar using a proboscis. Unlike **moths**, they are active during the day, have clubbed antennae, and fold their wings when resting.

caterpillar (*n*) the larva (↓) of a butterfly (↑), moth or other insect. It has a soft, round, worm-like body, usually covered in hairs, divided into segments. The head forms the first segment; there are two legs on each of the next three thoracic (p.154) segments, and a pair of structures called prolegs on each of the remaining ten abdominal (p.154) segments. Caterpillars feed on plants and leaves and some are serious pests (p.239).

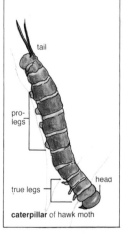

caterpillar of hawk moth

larva (*n. pl. larvae*) an immature form of many insects (p.140) and other invertebrates (p.139), unlike the adult form in appearance. It hatches from the fertilized egg and is unable to reproduce, e.g. tadpole, caterpillar. It changes into a pupa (↓) or adult during metamorphosis (↑). **larval** (*adj*).

pupa (*n. pl. pupae*) the resting stage between larva (↑) and adult in the life cycle of insects that undergo complete metamorphosis (↑). Pupae are often enclosed in protective cases, called cocoons (↓) and do not move or feed. The structure of the body alters substantially as the larva changes into the adult. **pupal** (*adj*).

cocoon (*n*) a cover made by many invertebrates to protect the developing larva (↑), pupa (↑) or egg, e.g. earthworms and some spiders produce cocoons to contain their eggs. The larvae of many insects spin a cocoon in which the pupae develop, e.g. silkworms make cocoons out of silk threads.

chrysalis (*n*) the pupal (↑) stage of butterflies, moths and certain other insects.

female

imago / adult

butterfly

adult lays eggs on the underside of the leaves of the larva's food plants

larva grows and changes its skin

egg (enlarged)

larva

chrysalis

pupa

insect pupates

egg hatches

imago (*n*) the adult stage of an insect after metamorphosis. It is sexually mature and can reproduce to form eggs which continue the life cycle.

moult (*v*) to shed an outer covering of skin, hair or feathers. It occurs periodically and may depend upon the growth rate of the individual, e.g. snakes moult their skins as they grow bigger, or upon seasonal changes, such as temperature, e.g. dogs moult hair in hot weather. In insects which undergo incomplete metamorphosis (↑), the outer covering of the larva splits open and the larva (↑) emerges with a new soft cover which gradually hardens in the air.

maggot (*n*) the soft, worm-like stage, or larva, of certain insects, in particular flies. Maggots do not have heads or legs.

social insects an insect which lives and works as a member of a large and highly organized group. Social insects are often divided into castes (↓), e.g. wasps and termites.

society (*n*) a group of animals that live and work together for mutual benefit. A group of social insects (↑) forms a society.

caste (*n*) in social insects, any group of insects that have a particular structure and function, and a particular job to do, e.g. honey bees are divided into three castes: queens (↓), workers (↓) and drones (↓).

queen (*n*) a female bee whose only duty is to lay eggs. In a hive (↓) of honey bees there is only one queen; she is larger than the other members of the hive.

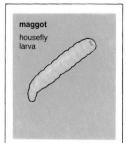

maggot
housefly
larva

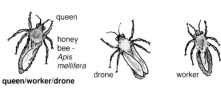

queen

honey
bee -
*Apis
mellifera*

drone

worker

queen/worker/drone

worker (*n*) a female social insect unable to reproduce. The workers are smaller than the queen and the drone, and perform all the work in the hive.

drone (*n*) a male bee whose duty is to fertilize the queen. Drones do no work.

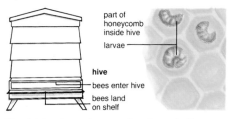

part of
honeycomb
inside hive

larvae

hive

bees enter hive

bees land
on shelf

hive (*n*) the place where a colony (group) of bees lives.

solitary insect an insect that lives alone, or as a member of a pair; it does not live in a large social group.

cell (*n*) a tiny unit of plant or animal life. It consists of protoplasm (p.150) enclosed in a membrane (p.150), and usually possesses a nucleus (p.150). Plant cells also have a cell wall. Cells absorb chemicals and process them to make the substances that they require for living. Some organisms are unicellular (one cell only), e.g. amoeba, whereas others contain many millions of cells, e.g. humans. There are many different types of cells, each having a particular function within the plant or animal, e.g brain cells; muscle cells. New cells are formed from an existing cell by division, or from existing cells by joining (fusion). **cellular** (*adj*).

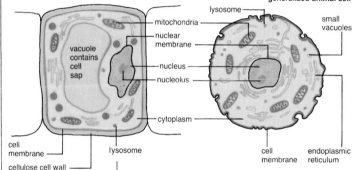

generalized plant cell

generalized animal cell

- lysosome
- mitochondria
- nuclear membrane
- nucleus
- nucleolus
- cytoplasm
- small vacuoles
- vacuole contains cell sap
- cell membrane
- lysosome
- cellulose cell wall
- cell membrane
- endoplasmic reticulum

structure (*n*) (1) the arrangement, or manner of organization, of the various constituent parts that form a whole cell, tissue, organ or organism, e.g. the structure of a plant cell; the structure of lung tissue; the structure of a human heart. (2) a complete part of an organism that has a particular function, e.g. an eye, a plant stem, and a tooth are all structures.

function (*n*) the specialized action of an organism, or part of an organism, performed in order to achieve a specific purpose, e.g. the function of plant roots is to take in water and soil nutrients; the function of the human heart is to pump blood round the body.

specialization (*n*) the adaption of an organism, part of an organism, or a cell in order to perform, or improve a particular function (↑), e.g. the specialization of cells like brain cells, or liver cells.

tissue (*n*) a group of many hundreds of plant or
animal cells (P.149) and intercellular matter which
together perform a specific function (p.149),
e.g. nervous tissue conducts impulses. Tissue is
composed of one or several types of cell,
e.g. nervous tissue consists of nerve cells and cells
that bind the tissue together.

cell wall a rigid wall of cellulose (↓) round the cell
membrane (↓) of a plant (p.139). It is made by the
protoplasm (↓), and is non-living. The cell wall gives
shape to a cell, and provides the firmness and
rigidity which allow a plant to stand upright.

membrane (*n*) a thin film of tissue covering,
separating or supporting a part of a living organism.
A **cell membrane** is a very thin, flexible membrane
found in all cells; it is composed of fat (p.165) and
protein (p.163). The cell membrane encloses the
contents of the cell, and controls the passage of
substances into and out of the cell. If the membrane
is damaged the cell dies.

permeable (*adj*) describes a substance through
which molecules (p.45) and ions (p.76) can pass. A
permeable substance may be impervious (↓).

impermeable (*adj*) describes a substance through
which molecules and ions cannot pass, i.e. it is not
permeable (↑).

impervious (*adj*) describes a substance through
which a fluid cannot pass. An impervious substance
may be permeable (↑) or impermeable (↑).

protoplasm (*n*) a clear, greyish, jelly-like material
that is the living matter of plant or animal cells. It
includes the nucleus (↓), cell membrane and
cytoplasm (↓). **protoplasmic** (*adj*).

nucleus² (*n*) a small, dense, membrane-bounded
body found in the cytoplasm (↓) of most plant and
animal cells. It contains chromosomes (p.152). The
nucleus controls the functions of the cell, such as
growth and reproduction. If the nucleus is
damaged, the cell dies.

cytoplasm (*n*) all the protoplasm (↑) of a cell, apart
from the nucleus (↑). Cytoplasm is clear and
jelly-like, and contains many different organelles
(p.153). It is divided into an inner layer, called
endoplasm (↓), and an outer layer, called ectoplasm
(↓). Cytoplasm is in a state of continuous movement.
cytoplasmic (*adj*).

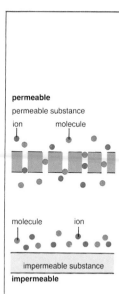

permeable
permeable substance

impermeable substance
impermeable

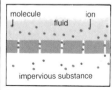

impervious substance
impervious

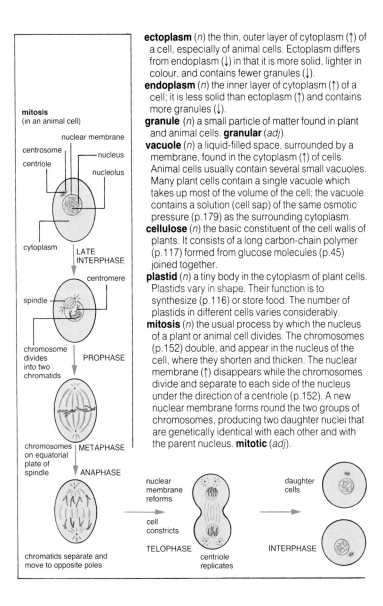

mitosis
(in an animal cell)

nuclear membrane

centrosome — nucleus

centriole

nucleolus

cytoplasm | LATE INTERPHASE

centromere

spindle

chromosome divides into two chromatids | PROPHASE

chromosomes on equatorial plate of spindle | METAPHASE

ANAPHASE

nuclear membrane reforms

daughter cells

cell constricts

chromatids separate and move to opposite poles

TELOPHASE

centriole replicates

INTERPHASE

ectoplasm (*n*) the thin, outer layer of cytoplasm (↑) of a cell, especially of animal cells. Ectoplasm differs from endoplasm (↓) in that it is more solid, lighter in colour, and contains fewer granules (↓).

endoplasm (*n*) the inner layer of cytoplasm (↑) of a cell; it is less solid than ectoplasm (↑) and contains more granules (↓).

granule (*n*) a small particle of matter found in plant and animal cells. **granular** (*adj*).

vacuole (*n*) a liquid-filled space, surrounded by a membrane, found in the cytoplasm (↑) of cells. Animal cells usually contain several small vacuoles. Many plant cells contain a single vacuole which takes up most of the volume of the cell; the vacuole contains a solution (cell sap) of the same osmotic pressure (p.179) as the surrounding cytoplasm.

cellulose (*n*) the basic constituent of the cell walls of plants. It consists of a long carbon-chain polymer (p.117) formed from glucose molecules (p.45) joined together.

plastid (*n*) a tiny body in the cytoplasm of plant cells. Plastids vary in shape. Their function is to synthesize (p.116) or store food. The number of plastids in different cells varies considerably.

mitosis (*n*) the usual process by which the nucleus of a plant or animal cell divides. The chromosomes (p.152) double, and appear in the nucleus of the cell, where they shorten and thicken. The nuclear membrane (↑) disappears while the chromosomes divide and separate to each side of the nucleus under the direction of a centriole (p.152). A new nuclear membrane forms round the two groups of chromosomes, producing two daughter nuclei that are genetically identical with each other and with the parent nucleus. **mitotic** (*adj*).

binary fission the asexual reproduction (p.225) of a cell by dividing into two cells of equal size. The nucleus divides into two nuclei, which then separate; the cytoplasm then divides and cell walls form around the two new cells.

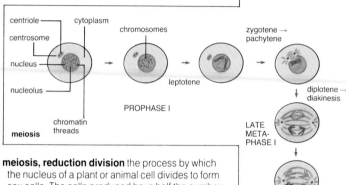

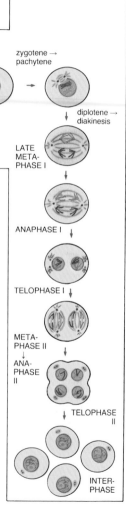

meiosis, reduction division the process by which the nucleus of a plant or animal cell divides to form sex cells. The cells produced have half the number of chromosomes (↓) of the parent cell, which means that when fusion (p.205) occurs during sexual reproduction (p.225), the cells produced have the normal number of chromosomes. During meiosis, chromosomes appear in pairs; each pair separates with one chromosome going to each side of the nucleus under the control of a centriole (↓). The two nuclei formed have half the number of chromosomes of the parent cell. The chromosomes then double, and a further mitotic (p.151) division occurs. The cytoplasm divides forming four sex cells with four new nuclei; each cell contains half the normal number of chromosomes.

centriole (*n*) a tiny, dense mass found just outside the nucleus (p.150) in most plant and animal cells. Before mitosis (p.151) the centriole divides into two.

polar body a minute cell, containing a nucleus and very little cytoplasm. It is formed during the formation of an egg-cell (a female sex cell). Polar bodies die.

chromosome (*n*) a microscopic, threadlike structure found in the nucleus of plant or animal cells. Chromosomes occur in pairs in all cells, apart from sex cells which contain only half the normal number

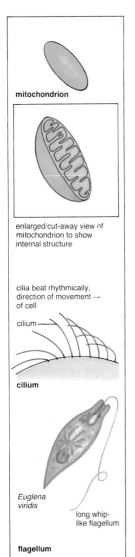

mitochondrion

enlarged/cut-away view of mitochondrion to show internal structure

cilia beat rhythmically, direction of movement → of cell

cilium

cilium

Euglena viridis

long whip-like flagellum

flagellum

of chromosomes. Each species has a constant number of chromosomes in each cell, e.g. humans have 46 chromosomes (23 chromosomes in sex cells). Chromosomes appear during cell division, and contain the genes which determine the characteristics of an organism, such as the colour and shape of a flower, or the colour of eyes and hair. The **sex chromosomes** determine the sex of an animal. A pair of X-chromosomes produces an animal of one sex; an X-chromosome paired with a Y-chromosome, or no other chromosome, produces an animal of the opposite sex. In mammals, the combination of an X and a Y chromosome produces a male; but if the sex chromosomes are XX the result is a female.

mitochondrion (*n. pl. mitochondria*) any one of a number of small, rod-like structures found in varying numbers in the cytoplasm of all cells. Mitochondria contain enzymes (p.161); their function is to use oxygen to release energy in aerobic respiration.

ribosome (*n*) a minute particle found in great numbers in the cytoplasm (p.150) of all cells. Ribosomes synthesize proteins that are used by the cell.

organelle (*n*) a part of a cell with a definite structure and function (p.149), e.g. plastids (p.151) and mitochondria (↑) are organelles.

cilium (*n. pl. cilia*) one of many short, hair-like projections which grow from the surface of certain plant or animal cells. Cilia are capable of beating rhythmically and thereby pushing fluid and particles past a cell, or propelling the cell through water. **ciliary** (*adj*).

flagellum (*n. pl. flagella*) a long, threadlike structure which has a beating, wave-like motion in water. The movement of the flagellum propels an organism. Flagella are larger than cilia (↑). Some unicellular organisms have only one or two flagella, while bacteria may have groups of several flagella called **tufts**.

pseudopodium (*n. pl. pseudopodia*) a portion of the cytoplasm (p.150) of a cell, extended as a temporary arm of irregular shape in some protozoa. Pseudopodia are formed to assist in locomotion and feeding, and are then taken back into the cell. Locomotion produced by the movement of pseudopodia is called **amoeboid movement**.

organ (*n*) in animals and plants, a structure made of specialized tissue which performs a particular function, e.g. in animals the heart is an organ whose function is to pump blood around the body.

system (*n*) the set of organs (↑) that together perform a particular function, e.g. the circulatory system conveys oxygen and digested food around the body; in humans the circulatory system consists of the heart, arteries, veins and capillaries.

viscera (*n*) the internal organs of an animal found in the abdomen (↓) and thorax (↓), e.g. heart, lungs, stomach, liver, kidneys and intestines.

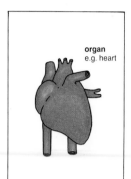

organ
e.g. heart

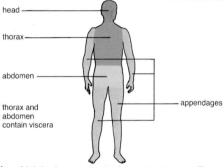

head

thorax

abdomen

thorax and
abdomen
contain viscera

appendages

head (*n*) the foremost part of an animal's body. The brain is located in the head.

thorax (*n*) in vertebrates (p.140), other than fish, the part of the body between the head and the abdomen (↓). It contains the heart and lungs, and in mammals is separated from the abdomen by the diaphragm (p.176) muscle. In insects it bears the legs and wings. **thoracic** (*adj*).

abdomen (*n*) (1) in vertebrates, the body cavity next to the thorax (↑) which contains the liver, kidneys and intestines. (2) the rear part of the body of an insect. **abdominal** (*adj*).

appendage (*n*) a projection from the main body of an animal, e.g. arm, leg, tail. **Paired appendages** are common, e.g. the antennae on the head of many arthropods. **appendicular** (*adj*).

fin (*n*) a flat, thin structure that projects from the body of fish. It consists of many small bones joined by a band of skin and is used for controlling the direction of movement, and for balancing in water.

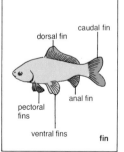

caudal fin

dorsal fin

anal fin

pectoral
fins

ventral fins

fin

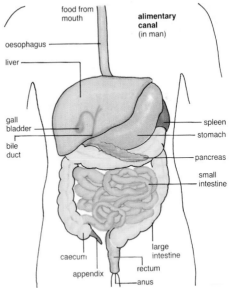

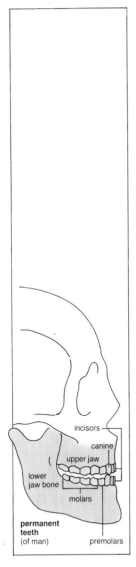

permanent
teeth
(of man)

alimentary canal the tube, in animals, through which
foodstuffs (p.166) pass, are digested and absorbed
by the body, and the waste eliminated. In all
animals, except very simple ones, such as
flatworms, it has two openings, the mouth where
food enters and the anus where unassimilated
material is expelled.

tooth (*n. pl. teeth*) (1) a small, sharp, hard structure
found in the mouth of vertebrates. Teeth are used
for biting and grinding food, and fighting and
seizing other animals. The tooth is divided into two
parts, the crown which is covered in enamel (p.156),
and the root which lies inside the gum. (2) any small
pointed part sticking out, e.g. a tooth on a saw, a
tooth on a leaf.

permanent teeth the second set of teeth which
replace the milk teeth in mammals. There are four
types: molars (p.156), premolars (p.156), canines
(p.156) and incisors (p.156), and the number and
size vary from species to species. The first set (milk
teeth) has fewer teeth, contains no molars, and
grows in young mammals.

dental (*adj*) of, or concerned with, the teeth, e.g. dental formula.

dentate (*adj*) (1) an animal possessing teeth. (2) a leaf with small pointed parts.

incisor (*n*) a tooth found at the front of the mouth in mammals. It is sharp and chisel-shaped, has one root and is used for gnawing, biting and cutting. In rodents the incisors grow in length continuously; they are worn down by use.

canine (*n*) a sharp, pointed tooth, usually found behind the incisors (↑) on either side of the upper and lower jaws. They are well-developed in carnivores and are used for holding prey and tearing or biting flesh. In rodents and some herbivores they are small or absent.

premolar (*n*) a tooth growing between the molars (↓) and canine (↑) teeth. It is similar in shape and function to a molar, but has only one or two roots; it is present in the first and second set of teeth of a mammal.

molar (*n*) a tooth growing at the back of the mouth behind the premolars (↑). It is similar to a premolar with a flat uneven surface used for grinding and crushing food. It has two or three roots and is present only in the permanent teeth of a mammal.

crown (*n*) the top part of a tooth visible above the gum. It is covered in enamel (↓).

enamel (*n*) a hard, white substance which covers the dentine (↓) on the crown (↑) of a tooth.

dentine (*n*) the hard tissue that forms most of the structure of a tooth. It lies under the enamel (↑) and is chemically similar to bone, but contains no cells.

pulp cavity the cavity beneath the dentine (↑) of a tooth. It contains dental pulp, nerves, blood vessels, tissues and cells producing dentine. Nerves and blood vessels enter the pulp through a narrow canal at the base of each root.

tongue (*n*) in vertebrates, a soft, fleshy, muscular structure, attached to the floor of the mouth. It is used for tasting, for helping to push food down the throat for swallowing, and in humans for making the sounds used in speech.

taste-buds (*n*) receptors found in groups on the tongue; they are sensitive to taste. Each taste-bud is sensitive to only one of four tastes: sweet, sour, bitter and salty.

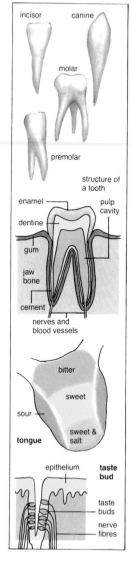

incisor canine

molar

premolar

structure of a tooth

enamel — pulp cavity

dentine

gum

jaw bone

cement

nerves and blood vessels

bitter

sweet

sour

sweet & salt

tongue

epithelium **taste bud**

taste buds

nerve fibres

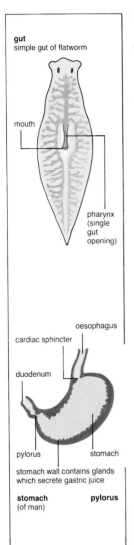

gut
simple gut of flatworm

mouth

pharynx
(single
gut
opening)

oesophagus

cardiac sphincter

duodenum

pylorus   stomach

stomach wall contains glands
which secrete gastric juice

stomach   **pylorus**
(of man)

gut (*n*) in animals, the part of the tube through which food passes, where food is digested by digestive juices. The soluble food substances are absorbed through the gut and used by the body. In simple animals, such as flatworms, it has only one opening. In mammals, it is a system of many organs, glands and tubes.

oesophagus (*n*) in vertebrates (p.140), the muscular tube between the pharynx and the stomach (↓). Food passes along it by peristalsis (↓). In birds the oesophagus contains the crop.

bolus (*n*) a ball of chewed food and saliva which is passed down the oesophagus (↑) by peristalsis (↓).

peristalsis (*n*) the wave-like contractions of the muscles in the wall of the alimentary canal (p.155) which push the contents from one end to the other. The contractions also mix the food with digestive juices. **peristaltic** (*adj*).

stomach (*n*) the bag-like structure which forms part of the alimentary canal (p.155). In vertebrates, except birds, it follows the oesophagus (↑). It has muscular walls which press and churn the food, and gastric juices are secreted from cells in the stomach lining; these juices carry out digestive processes. In some animals the stomach is in separate sections, e.g. ruminants have several chambers; birds have a crop and gizzard.

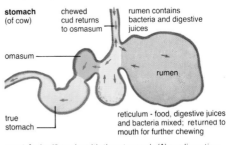

stomach
(of cow)

chewed
cud returns
to osmasum

rumen contains
bacteria and digestive
juices

omasum

rumen

true
stomach

reticulum - food, digestive juices
and bacteria mixed; returned to
mouth for further chewing

gastric (*adj*) to do with the stomach (↑) or digestive processes.

pylorus (*n*) the opening from the stomach (↑) into the duodenum (p.158). It is controlled by a sphincter (p.183) muscle which prevents food from passing through until the digestive processes in the stomach have been completed.

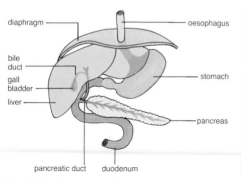

diaphragm

oesophagus

bile duct

gall bladder

stomach

liver

pancreas

pancreatic duct   duodenum

liver (*n*) the largest glandular organ in vertebrates. It has many functions which include the production of bile, deamination of amino acids, storage of glycogen, minerals and vitamins, and destruction of foreign chemicals. These metabolic activities are the main source of the heat which maintains the mammalian body temperature.

hepatic (*adj*) to do with the liver (↑), e.g. the hepatic duct takes bile (p.162), from the liver, to the gall bladder (↓).

gall bladder a small gland present in many vertebrates. It lies between the lobes of the liver (↑). The liver continuously secretes bile which is passed to the gall bladder via ducts, where it is stored. The bile is expelled into the intestine (↓) when food, especially fat, is present.

bile duct a duct which carries bile from the liver to the duodenum (↓).

pancreas (*n*) in all vertebrates, except some fish, a gland found near the duodenum (↓). It produces pancreatic juices, an alkaline mixture of enzymes (p.161) mainly trypsinogen (p.163), lipase (p.163), amylase (p.161) and maltase (p.163) which pass via the pancreatic duct into the duodenum. The pancreas also produces the hormone insulin from a special group of cells called the islets of Langerhans.

duodenum (*n*) the first part of the small intestine (↓) in mammals, between the stomach (p.157) and the ileum. The bile duct (↑) and pancreatic duct open into it. Food is digested by secretions from these ducts and from the walls of the duodenum.

gall bladder
a function of bile

large globule of fat

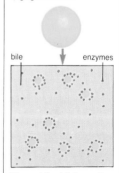

bile   enzymes

bile emulsifies fat; the small globules present a large surface area for enzymes to break it down to fatty acids and glycerol

ileum (*n*) the last part of the small intestine (↓) which leads into the large intestine (↓). The absorption of digested food is completed here.

small intestine the long muscular tube in mammals, birds and reptiles where the digestion of food is completed (except in herbivores) and the absorption of food takes place. The internal surface area of the small intestine is increased by villi (↓) which absorb the soluble food products.

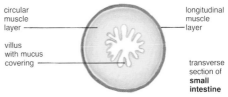

circular muscle layer

longitudinal muscle layer

villus with mucus covering

transverse section of **small intestine**

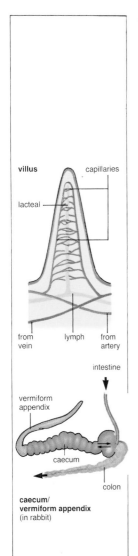

villus

capillaries

lacteal

from vein   lymph   from artery

intestine

vermiform appendix

caecum

colon

caecum/ vermiform appendix (in rabbit)

villus (*n. pl. villi*) a finger-like projection. Villi line the walls of the small intestine. They are present in very large numbers and greatly increase the absorptive surface of the intestine. Each contains a lacteal into which fats pass, surrounded by blood capillaries that absorb amino acids (p.163) and also the monosaccharides (p.164).

intestine (*n*) in vertebrates, the long, tube-like part of the alimentary canal (p.155) between the stomach (p.157) and rectum (p.160). It is concerned with the digestion and absorption of food. In man, it is about 3½m in length and lies coiled in the abdominal cavity. Food is passed along by muscular contractions called peristalsis (p.157). In mammals, birds and reptiles it is divided into the small intestine (↑) and large intestine (↓).

large intestine the second part of the intestine. It is a broad muscular tube and is shorter in length than the small intestine (↑). It absorbs water and mineral salts from the material passing through it, forming faeces (p.160) from the rejected material.

caecum (*n*) a blind-ended branch of the gut (p.157) of vertebrates. In herbivores the caecum is large and helps in the digestion of cellulose. It is much smaller in carnivores, and in humans it has no function.

vermiform appendix a small, finger-like tube, closed at one end, which leads from the caecum (↑). It contains lymphoid (p.173) tissue.

colon (*n*) the first part of the large intestine (p.159) in which water from undigested food is absorbed. It is wider than the rectum (↓) into which it leads.

rectum (*n*) the short tube which forms the second part of the large intestine (p.159). It stores faeces (↓) which is expelled through the anus (↓) or cloaca (↓).

roughage (*n*) the indigestible part of food. It is necessary to encourage intestinal movement and gives bulk to faeces. In man's diet it is usually in the form of cellulose.

faeces (*n.pl.*) the undigested waste from food which is passed through the alimentary canal (p.155). It is expelled from the anus (↓) in mammals and the cloaca (↓) in other vertebrates.

defaecate (*v*) to expel faeces (↑) from the anus (↓) or cloaca (↓).

anus (*n*) the terminal opening of the alimentary canal in mammals. Faeces are expelled through it. It is closed by the anal sphincter (p.183), a circular muscle.

cloaca (*n*) the posterior end of the alimentary canal of vertebrates, except mammals, and some invertebrates. The alimentary tract, kidney and reproductive ducts enter into it. It has an opening to the exterior.

metabolism (*n*) the chemical processes occurring within an organism, or part of it. They involve the decomposition of organic compounds (catabolism) and the release of energy, e.g. digestion, and the synthesis of new compounds from simple molecules (anabolism), e.g. the synthesis of proteins from amino acids.

digestion (*n*) the chemical decomposition of foodstuffs into substances that can be absorbed by an organism for metabolic purposes. In most animals this takes place in the gut (p.157) and is carried out by enzymes (↓).

absorb (*v*) for a liquid or solid to take into itself another substance, usually a liquid or gas. This process can be by diffusion, osmosis or capillary attraction, e.g. blood absorbs oxygen from the lungs, plants absorb water through their roots.

assimilation (*n*) the utilization of simple molecules, e.g. digested foodstuffs such as amino acids, into the complex constituents of the organism, such as proteins. **assimilate** (*v*).

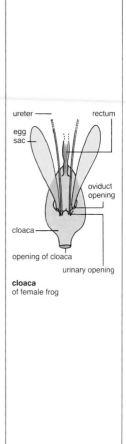

ureter — rectum

egg sac

oviduct opening

cloaca —

opening of cloaca

urinary opening

cloaca
of female frog

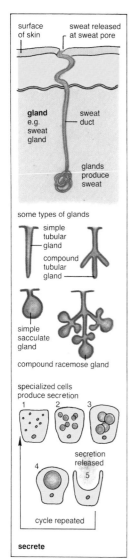

some types of glands

simple
tubular
gland

compound
tubular
gland

simple
sacculate
gland

compound racemose gland

specialized cells
produce secretion

secretion
released

cycle repeated

secrete

gland (*n*) in animals, an organ whose main function is to form specific chemical substances for secretion. There are two types, **exocrine** glands which secrete their solutions through a duct (↓) either internally, e.g. digestive glands onto the internal surface of the gut, or externally onto the outer surface of the skin, e.g. sweat glands. The **endocrine** glands (p.204) have no ducts and secrete directly into the blood stream. **glandular** (*adj*).

secrete (*v*) to discharge a material made by a cell for use in an organism. Secretion is an activity of most cells and is specialized in gland cells. **secretion** (*n*).

duct (*n*) a tube or channel in animals and plants which carries fluid from one system to another.

juice (*n*) a liquid secretion from a digestive gland (↑). It contains enzymes, e.g. pancreatic juice from the pancreas (p.158); gastric juice from the stomach (p.157). In plants, the fruit juice obtained by crushing fruit.

saliva (*n*) a secretion (↑) from the salivary glands in the mouth. In land vertebrates it contains mucus and sometimes ptyalin (↓). It is secreted as a reflex response to the presence of food and lubricates the food for swallowing. In insects it is the fluid secreted onto the mouth-parts and often contains digestive enzymes. In bloodsuckers saliva contains an anticoagulant.

enzyme (*n*) a protein which acts as a catalyst (p.108), converting one or more substances, the substrates (p.144), into other substances, the products. Enzymes are produced by living cells of bacteria, plants and animals. Most enzymes activate only one kind of substrate so an organism needs to produce a large number of different enzymes for all the different reactions of metabolism, e.g. digestion of food. Enzymes are readily destroyed by high temperatures or by chemical substances. They work best under specific conditions, e.g. suitable pH and temperature.

ptyalin (*n*) an enzyme (↑) found in the saliva (↑) of some mammals, e.g. man. It changes starch into maltose (p.164).

amylase (*n*) a group of enzymes found in plants and the saliva (↑) and pancreatic juice (p.162) of animals. It acts on starch or glycogen (p.165) to produce maltose and glucose (p.164).

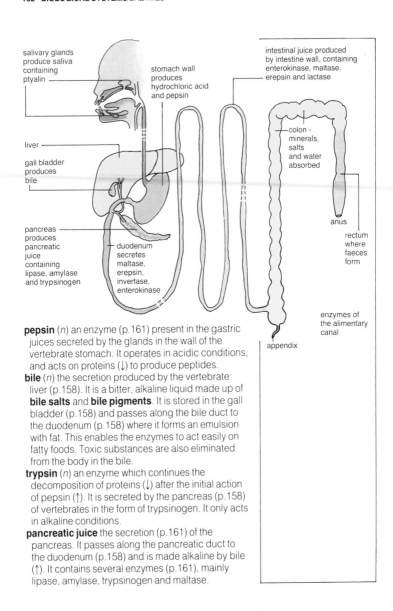

salivary glands produce saliva containing ptyalin

stomach wall produces hydrochloric acid and pepsin

intestinal juice produced by intestine wall, containing enterokinase, maltase, erepsin and lactase

liver

gall bladder produces bile

colon - minerals, salts and water absorbed

pancreas produces pancreatic juice containing lipase, amylase and trypsinogen

duodenum secretes maltase, erepsin, invertase, enterokinase

anus

rectum where faeces form

enzymes of the alimentary canal

appendix

pepsin (*n*) an enzyme (p.161) present in the gastric juices secreted by the glands in the wall of the vertebrate stomach. It operates in acidic conditions, and acts on proteins (↓) to produce peptides.

bile (*n*) the secretion produced by the vertebrate liver (p.158). It is a bitter, alkaline liquid made up of **bile salts** and **bile pigments**. It is stored in the gall bladder (p.158) and passes along the bile duct to the duodenum (p.158) where it forms an emulsion with fat. This enables the enzymes to act easily on fatty foods. Toxic substances are also eliminated from the body in the bile.

trypsin (*n*) an enzyme which continues the decomposition of proteins (↓) after the initial action of pepsin (↑). It is secreted by the pancreas (p.158) of vertebrates in the form of trypsinogen. It only acts in alkaline conditions.

pancreatic juice the secretion (p.161) of the pancreas. It passes along the pancreatic duct to the duodenum (p.158) and is made alkaline by bile (↑). It contains several enzymes (p.161), mainly lipase, amylase, trypsinogen and maltase.

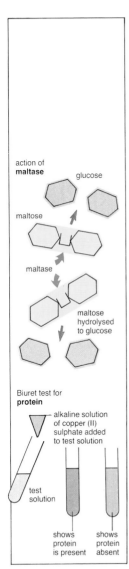

action of
maltase

glucose

maltose

maltase

maltose
hydrolysed
to glucose

Biuret test for
protein

alkaline solution
of copper (II)
sulphate added
to test solution

test
solution

shows
protein
is present

shows
protein
absent

trypsinogen (*n*) the inactive form in which trypsin (↑) is secreted by the pancreas. It is activated by the enzyme enterokinase (↓).

lipase (*n*) an enzyme present in pancreatic juice (↑). It hydrolyses fats into glycerol and organic acids. Lipase only acts in the alkaline conditions formed by bile (↑).

amylopsin (*n*) an amylase (p.161) produced by the pancreas (p.158) of vertebrates and found in pancreatic juice (↑).

intestinal juice the juice secreted by the glands in the wall of the vertebrate duodenum. It contains digestive enzymes, e.g. maltase, erepsin, invertase and enterokinase (↓) which complete the digestion of protein (↓) to amino acids (↓) and starch (p.165) to glucose (p.164).

invertase (*n*) an enzyme produced by plants and animals. In man it is present in intestinal juice (↑). Invertase hydrolyses sucrose into glucose and fructose.

enterokinase (*n*) an enzyme (p.161) secreted by the glands in the small intestine (p.159). It acts on trypsinogen (↑) converting it to its active form of trypsin (↑).

maltase (*n*) an enzyme (p.161) found in the intestinal juice (↑) and pancreatic juice (↑) of man. It hydrolyses maltose into glucose.

erepsin (*n*) a mixture of enzymes (p.161) present in intestinal juice (↑) which complete the decomposition of protein (↓) to amino acids (↓) after the initial action of pepsin and trypsin (↑).

protein (*n*) a complex compound formed from a chain of hundreds or thousands of different amino acid (↓) molecules. Proteins are present in all living cells, both structurally and as enzymes (p.161).

polypeptide (*n*) a chain compound consisting of many molecules of amino acids, but producing chains much shorter than those in proteins. Polypeptides are formed by the decomposition of proteins (↑) by digestive juices. They are further decomposed to form amino acids (↓).

amino acid an organic compound possessing both an acid group ($-COOH$) and an amino group of atoms ($-NH_2$) in a molecule. Amino acids are fundamental constituents of living matter as they are combined together to form protein molecules.

carbohydrate (*n*) a complex compound consisting of carbon, hydrogen and oxygen; the general formula is $C_x(H_2O)_y$. The three main types of carbohydrates are sugars (↓), starch (↓), and cellulose (p.151). Carbohydrates play an essential part in the metabolism (p.160) of all living organisms. They are present in much larger amounts in plants than in animals.

sugar (*n*) the simplest kind of carbohydrate (↑). It has a crystalline structure, a sweet taste and is soluble in water. Various sugars are produced by plants in photosynthesis (p.223).

glucose (*n*) a monosaccharide (↓) with the formula $C_6H_{12}O_6$. In plants it is produced by photosynthesis (p.223) and stored as starch (↓). In animals it is obtained by the digestion of carbohydrates and is stored in the liver (p.158) as glycogen (↓).

fructose (*n*) a monosaccharide (↓) with the formula $C_6H_{12}O_6$. It is found in many plants, particularly fruits.

monosaccharide (*n*) the simplest group of sugars (↑), containing either 5 or 6 carbon atoms. When decomposed they cease to have the properties of sugar. Glucose and fructose both have the molecular formula of a monosaccharide with 6 carbon atoms, i.e. $C_6H_{12}O_6$, but they have different properties. Monosaccharides are able to pass through most cell membranes.

maltose (*n*) a disaccharide (↓) with the formula $C_{12}H_{12}O_{11}$. It is formed as a result of starch hydrolysis during digestion (p.160). One molecule of maltose is a compound of two molecules of glucose which are split by the action of the enzyme maltase (p.163). Maltose is also found in some germinating (p.232) seeds.

disaccharide (*n*) a sugar, formed by the combination of two monosaccharide (↑) molecules which can be either the same or different monosaccharides. It has 12 carbon atoms and a formula $C_{12}H_{22}O_{11}$. Disaccharide molecules can be hydrolysed into two monosaccharide molecules. Disaccharides are soluble in water, but are unable to pass through cell membranes.

sucrose, cane sugar a disaccharide (↑) composed of a glucose (↑) and a fructose (↑) molecule. It is split by the enzyme (p.161) sucrase. It is widely found in plants, but not in animals.

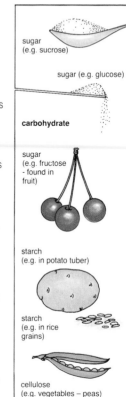

sugar
(e.g. sucrose)

sugar (e.g. glucose)

carbohydrate

sugar
(e.g. fructose
- found in
fruit)

starch
(e.g. in potato tuber)

starch
(e.g. in rice
grains)

cellulose
(e.g. vegetables – peas)

monosaccharide

CH₂OH

e.g. structure of glucose
– a monosaccharide

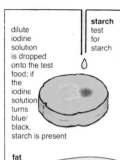

dilute iodine solution is dropped onto the test food; if the iodine solution turns blue/black, starch is present

starch test for starch

fat

test for fats:
the test food is smeared on to a piece of paper; if a translucent area is left behind, fats are present

vitamin

vitamin	source
A	carrots, liver, butter and green vegetables
B₁	yeast, meat, milk, peas, beans, whole-meal bread
B₂	yeast, lean meat, peas, eggs, liver, milk and green vegetables
C	rose 'hips', grapefruits, lemons, oranges, green vegetables, tomatoes and blackcurrants
D	vegetable oil, eggs, fish liver oil
E	eggs, liver, wheat germ, green vegetables

polysaccharide (*n*) a carbohydrate produced by the combination of many hundreds of monosaccharide (↑) molecules, usually of the same type. Polysaccharides form large, often fibrous molecules. Both plants and animals store energy in the form of polysaccharides. When hydrolysed, they produce disaccharides (↑) or monosaccharides (↑).

starch (*n*) a polysaccharide (↑) formed from the chemical combination of many glucose (↑) molecules (p.45). It is a white, non-crystalline substance, insoluble in water. In plants, carbohydrates (↑) are stored as starch and when hydrolysed starch forms glucose.

glycogen (*n*) a partially soluble polysaccharide formed from the chemical combination of many glucose (↑) molecules. In animals and some fungi (p.142), carbohydrates are stored as glycogen. In vertebrates it is mainly produced and stored in the liver (p.158) and muscles. When hydrolysed it forms glucose.

fat (*n*) an organic substance which is an ester (p.116) of fatty acids (↓) and glycerol (↓); it contains the elements (p.44) carbon, hydrogen and oxygen. True fats are solid at room temperature, e.g. butter. Oils are liquid at room temperature. Although known as fats, other compounds which can be dissolved in hot ethanol are not true fats. **fatty** (*adj*).

fatty acid an organic acid present in plants and animals that usually has an even number of carbon atoms arranged in a straight chain, with a carboxyl group of atoms ($-COOH$) at the end. Acids commonly found in plants are oleic acid, palmitic acid and stearic acid.

glycerol (*n*) a sugar alcohol (p.116) with the formula $C_3H_6O_3$; it possesses three hydroxyl groups. It is an odourless, colourless, sweet, viscous liquid.

vitamin (*n*) a complex, organic substance required in small amounts by most organisms. It plays an essential role in the metabolism of an organism, and each organism has its own specific requirements for vitamins. Most vitamins are obtained from food, although occasionally some organisms can synthesize a particular vitamin, e.g. vitamin D in man. The table shows the main food sources of some vitamins.

diet (*n*) the typical food intake of an animal, i.e. the type and quality of the foods consumed, and their relative quantities.

balanced diet a diet which supplies an animal with the optimum amount of carbohydrates and fats (to provide energy), proteins (to replace amino acids), and vitamins, mineral salts, and roughage (p.160). A diet which is deficient in nutrients (↓), or provides too much of a particular nutrient, is called an **unbalanced diet**.

nutrition (*n*) the process by which an organism takes in and absorbs food to replace cellular material and promote growth. Nutrition can be divided into four stages; ingestion, digestion (p.160), absorption (p.39) and assimilation (p.160). **nutritious** (*adj*), **nutritional** (*adj*).

nutrient (*n*) any substance that can be used in the nutrition of a specific organism, e.g. mineral salts are nutrients for plants; vitamins are nutrients for animals.

foodstuff (*n*) a chemical classification of food consumed by animals. Foodstuffs are carbohydrates (p.164), proteins (p.163) and fats (p.165); they help repair cellular matter, provide energy, and aid growth.

calorie (*n*) a former measure of the heat or energy value of foodstuffs. A kilocalorie, written Calorie, is equal to 1000 calories. The calorie is now replaced by the joule (p.13).

energy value the heat energy made available by the complete combustion of a stated weight of food. It is given in joules per kilogram, or Calories (↑) per lb.

calorific value an alternative term for **energy value** (↑). *See* **calorie**.

food chain a set of organisms which depend directly on one another for food energy. The chain begins with a green plant which obtains its food from inorganic compounds. The plant is eaten by a herbivore (p.241) or an omnivore (p.241), which is in turn eaten by a carnivore (p.241) or an omnivore. A simple food chain is: grass—cow—human.

food web a number of linked food chains (↑). Many organisms feed on more than one organism, and are in turn eaten by more than one organism. The various food chains are thus linked together to form a food web.

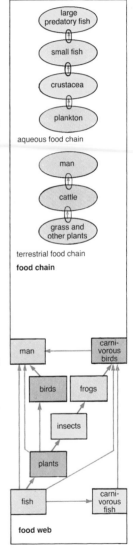

aqueous food chain

terrestrial food chain

food chain

food web

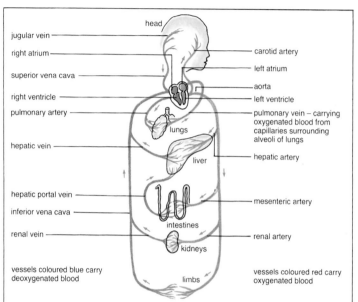

jugular vein

right atrium

superior vena cava

right ventricle

pulmonary artery

hepatic vein

hepatic portal vein

inferior vena cava

renal vein

head

carotid artery

left atrium

aorta

left ventricle

pulmonary vein – carrying oxygenated blood from capillaries surrounding alveoli of lungs

lungs

hepatic artery

liver

mesenteric artery

intestines

renal artery

kidneys

vessels coloured blue carry deoxygenated blood

vessels coloured red carry oxygenated blood

limbs

circulatory system
blood circulating system of *homo sapiens*

heart pumps blood round a complete double circulation

circulatory system the blood vascular system (↓) of an animal, especially vertebrates. It is made up of tubes, e.g. arteries (p.169), veins (p.169) and capillaries (p.170) and has a major organ, the heart (p.168) in vertebrates, which is responsible for pumping the liquid around the system. It carries in the liquid dissolved substances required for the metabolism of the organism.

vascular system a system, in vertebrates, of vessels to conduct fluids. The blood system conducts blood and the lymphatic system conducts lymph. The vascular system in plants conducts water, dissolved mineral salts and synthesized food materials.

vessel (*n*) a tube through which liquids are conducted round an organism, e.g. lymphatic vessels conduct lymph (p.173).

sinus (*n*) a small cavity in a tissue. Unlike a vessel (↑), it has a variable diameter, e.g. lymphatic sinus. Blood sinuses are found in the circulatory system of some animals, particularly invertebrates.

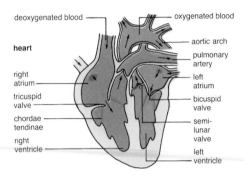

deoxygenated blood ——
oxygenated blood

heart

— aortic arch
— pulmonary artery

right atrium
tricuspid valve
chordae tendinae
right ventricle

left atrium
bicuspid valve
semi-lunar valve
left ventricle

heart (*n*) the muscular, hollow organ of an animal whose function is to pump blood around the circulatory system (p.167). The heart possesses valves to maintain a one-directional blood flow, pumped by contractions of the muscular walls. The vertebrate heart is divided into chambers, atria (↓) and ventricles (↓). Mammals and birds possess two atria and two ventricles; amphibians have two atria and one ventricle; fish have one atrium and one ventricle.

pacemaker (*n*) in vertebrates, a group of muscle cells in the heart where the nervous impulse starting the contraction of the atrium is sent out. The pacemaker is situated in the wall of the right atrium of mammals and birds.

auricle (*n*) an alternative name for **atria**, generally used for animals other than humans.

atrium (*n. pl. atria*) one of the chambers of the heart. It receives blood from the veins and pumps it into a ventricle (↓). The muscular wall is thinner and less powerful than that of a ventricle. In vertebrates, other than fish, oxygenated blood is received from the lungs into the left atrium, and deoxygenated blood received from the rest of the body passes into the right atrium.

ventricle (*n*) one of the chambers of the heart (↑). It has thick muscular walls for pumping blood round the body. In mammals the right ventricle receives deoxygenated blood from the right atrium and pumps it to the lungs. The left ventricle receives oxygenated blood from the left atrium and pumps it to the rest of the body.

pacemaker

pacemaker (sinu-atrial node)

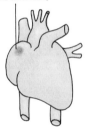

artificial pacemaker implanted in the body (sometimes under the arm) when the sinu-atrial node no longer functions properly

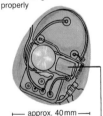

⊢— approx. 40mm —⊣

nuclear-powered battery ——

systole (*n*) the phase of heart beat in vertebrates when the cardiac muscle (p.188) contracts and blood is squeezed into the arteries. Atria contract before ventricles.

diastole (*n*) the phase of heart beat in vertebrates when the cardiac muscle (p.188) relaxes and the heart refills with blood from the veins.

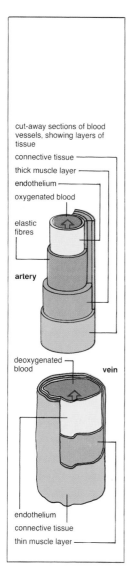

cut-away sections of blood vessels, showing layers of tissue

connective tissue

thick muscle layer

endothelium

oxygenated blood

elastic fibres

artery

deoxygenated blood   **vein**

endothelium

connective tissue

thin muscle layer

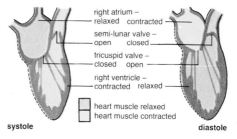

right atrium – relaxed   contracted

semi-lunar valve – open   closed

tricuspid valve – closed   open

right ventricle – contracted   relaxed

heart muscle relaxed
heart muscle contracted

systole   **diastole**

artery (*n*) a thick, muscular-walled vessel which carries blood from the heart (↑). The large aorta (↓) artery leaves the heart and branches into much smaller arteries which extend to every part of the body.

aorta (*n*) in mammals, including man, the main artery which carries oxygenated blood from the heart (↑) to every part of the body, except the lungs. In man, blood is passed through the aorta at about 4 litres per minute. Many arteries branch from the aorta.

arteriole (*n*) a small artery (↑) of vertebrates. The walls are composed of smooth muscle, and are controlled by the autonomic nervous system. This controls the blood supply to the capillaries (p.170).

vein (*n*) a blood vessel (p.167) which carries blood from the tissues (p.150) to the heart (↑). Veins have thinner walls and larger internal diameters than arteries (↑). They have valves which prevent the blood flowing backwards, ensuring it flows only to the heart.

vena cava the main vein, in vertebrates, except fish, which collects deoxygenated blood from all parts of the body, except the lungs, and passes it to the right atrium (↑) of the heart. The vein has two parts, one part collects blood from the hind limbs and trunk, the other part brings blood from the forelimbs and head.

portal system in vertebrates, the system of veins which carries blood from one capillary network (↓) to another, e.g. the renal portal system in fish, amphibians and reptiles conducts blood from the hind limbs to the kidneys (p.182).

capillary (*n*) a very small blood vessel (p.167). Capillaries receive blood from arterioles (p.169) and pass it to small veins (P.169). Capillaries are very numerous and form a network which spreads through nearly all tissues. They have thin walls which are permeable to water, dissolved oxygen and carbon dioxide, glucose, amino acids and inorganic ions.

capillary network the system of capillaries (↑) which spreads through tissues. The vessels branch and rejoin, penetrating all tissues.

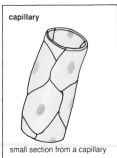

capillary

small section from a capillary

capillary network

venule

arteriole

artery

capillary network

vein

blood (*n*) the liquid carried by blood vessels (p.167) of an animal. In vertebrates it is passed round a circulatory system (p.167) by the pumping action of the heart. Blood is composed of liquid plasma (↓), containing white and red blood cells (↓), and platelets (only present in mammalian blood). It transports dissolved respiratory gases, products of digestion and excretion, hormones and mineral salts, and helps control the body temperature.

plasma (*n*) the clear, almost colourless liquid part of vertebrate blood after all the blood cells have been removed. Plasma clots as easily as whole blood.

corpuscle (*n*) (1) a cell in a fluid, e.g. a red blood corpuscle is a red blood cell. (2) a very small piece of a substance, a particle.

red blood cell a disc-shaped cell present in vertebrate blood. It contains haemoglobin (↓) which is responsible for the red colour. Red blood cells transport oxygen from the lungs to all parts of the body. They are easily distorted to enable them to pass through narrow capillaries (↑). They are incapable of movement on their own. In mammals they have no nucleus, but consist of an elastic membrane and cytoplasm. Human beings have

constituents of
plasma

| soluble food substances (sugars, amino acids and tiny fat droplets) |
| soluble waste products |
| soluble blood proteins including fibrinogen |
| mineral salts (mostly sodium and chloride ions) |
| vitamins, hormones and antibodies |

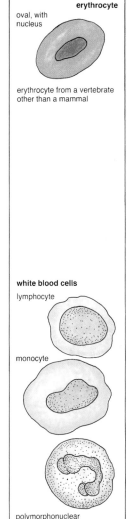

erythrocyte

oval, with nucleus

erythrocyte from a vertebrate other than a mammal

white blood cells

lymphocyte

monocyte

polymorphonuclear leucocyte

about five million red blood cells per 1 mm³. They are formed in red bone marrow and have a relatively short life, approximately four months in man.

erythrocyte (*n*) an alternative name for a **red blood cell** (↑).

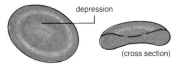

depression

erythrocyte from a mammal – round, no nucleus and biconcave

(cross section)

haemoglobin (*n*) a red pigment found in the red blood cells of vertebrates and the plasma of some invertebrates, e.g. earthworms. It readily combines with oxygen from the lungs and forms **oxyhaemoglobin** which is scarlet. The oxygen is transported to the body tissues where the oxyhaemoglobin decomposes to release oxygen. The deoxygenated haemoglobin is bluish-red in colour. Each animal species has a different type of haemoglobin, and in mammals that of the foetus is different from that of the adult.

white blood cell a colourless cell present in the blood and lymph of most vertebrates. It has a nucleus and is a polymorph, monocyte or lymphocyte. Unlike red blood cells (↑), white cells are motile (able to move about). They are important in the body's defence against infection. In man there are about 8000 white cells per 1 mm³ of blood.

leucocyte (*n*) an alternative name for **white blood cells**.

phagocyte (*n*) a cell that can engulf foreign bodies into its own cytoplasm and ingest them. This term is particularly applied to the polymorphs and monocytes present in blood. Phagocytes are an important defence mechanism against invading bacteria.

phagocytosis (*n*) the process of engulfing, or the ingestion of foreign bodies by phagocytes (↑).

lymphocyte (*n*) a type of vertebrate white blood cell (↑). It has a large nucleus and little cytoplasm. Lymphocytes are produced in lymphoid tissue, they are non-phagocytic (↑), show amoeboid movement and produce antibodies (p.216) in the blood. They make up about 25% of the white cells in human blood.

platelet (*n*) a very small fragment of a cell from red bone marrow found only in mammalian blood. In man there are about 250 000 per mm³ of blood. Platelets initiate the process of blood clotting (↓).

thrombin (*n*) an enzyme formed from the inactive blood-protein prothrombin by the activator thrombokinase. It converts the soluble blood-protein fibrinogen into insoluble fibrin (↓) during blood clotting.

fibrin (*n*) an insoluble protein that forms fibrous threads which tangle together to produce a mesh. Blood cells trapped in the mesh form a clot. Fibrin is produced from fibrinogen by the action of the enzyme thrombin (↑).

clot (*n*) a solid mass produced by coagulation in a liquid. A blood clot is formed when red blood cells (p.170) are trapped in the mesh produced by fibrin (↑). The clot prevents blood escaping from the injured blood vessels (p.167) and bacteria from entering the wound.

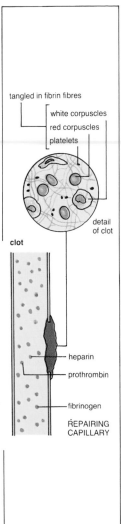

tangled in fibrin fibres

white corpuscles
red corpuscles
platelets

detail of clot

clot

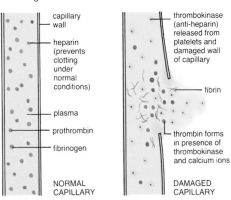

capillary wall

heparin (prevents clotting under normal conditions)

plasma

prothrombin

fibrinogen

NORMAL CAPILLARY

thrombokinase (anti-heparin) released from platelets and damaged wall of capillary

fibrin

thrombin forms in presence of thrombokinase and calcium ions

DAMAGED CAPILLARY

heparin

prothrombin

fibrinogen

REPAIRING CAPILLARY

serum (*n*) a yellowish liquid which separates from blood (p.170) or plasma when it clots. Its composition is the same as that of blood plasma, but it lacks the clotting constituents.

tissue fluid the fluid which surrounds all the cells of an animal. Tissue fluid diffuses through the walls of blood capillaries supplying tissue cells with oxygen and digested food substances. It removes carbon dioxide and excretory products from the cells.

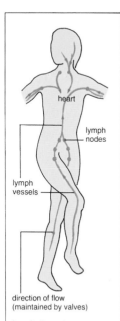

heart

lymph nodes

lymph vessels

direction of flow
(maintained by valves)

lymphatic system

lymph (*n*) a colourless fluid present in the lymphatic system (↓) of vertebrates. It contains less white cells and blood proteins than blood, has no red blood cells (p.170) but more lymphocytes. It is obtained from blood by filtration through capillary walls.

lymphatic system a system of thin-walled tubes in vertebrates that form a network for conducting lymph (↑). The system starts with lymph capillaries found in most tissues. Lymph capillaries are more permeable than blood capillaries, and bacteria are able to pass into the lymphatic system and are destroyed in lymph nodes (↓). Lymph capillaries join to form larger tubes, lymphatics (↓), which in turn join to form a lymph vessel which conducts lymph to a main vein near the heart. Lymphatics have valves similar to those in veins which maintain a one-directional flow.

lymphatic (*n*) a vessel which conducts lymph (↑) from the lymph capillaries.

lymph node a structure found in the lymph vessels of mammals; in birds they are not so well developed. Lymph nodes consist of lymphoid tissue. They produce lymphocytes (p.171) which remove bacteria from lymph and filter out foreign bodies.

spleen (*n*) in vertebrates, an organ which produces lymphocytes, removes red blood cells (p.170) from the blood and destroys them; it also stores and supplies red blood cells to the blood. The spleen is composed of lymphoid tissue, is situated close to the stomach and is connected to the circulatory system.

blood group the classification of human blood types into groups whose blood can be mixed without agglutination. The four main groups are: A, B, AB and O.

rhesus factor, Rh-factor an antigen (p.216) found on the red blood cells of the rhesus monkey and most humans. Individuals possessing the antigen are rhesus positive, and those lacking the antigen are rhesus negative. The negative individuals do not normally have antibodies (p.216) against the antigen. The antibody may be acquired by a blood transfusion (p.174), or as a result of a pregnancy involving a Rh positive foetus. Damage may then be caused, in any subsequent pregnancies, to a Rh positive foetus.

agglutination (*n*) the clumping together of cells. Red blood cells (p.170) clump together when blood from different groups is mixed.

transfusion (*n*) the process of transferring blood from one human being to another. The donor's (↓) blood must be compatible (↓) with the recipient's.

compatible (*adj*) of blood, being able to receive a transfusion (↑) without an adverse reaction. Red blood cells can have antigens (p.216), known as A and B. Group A red blood cells have antigen A, group B have antigen B, and group AB have both A and B antigens. Group O possesses no antigens. Plasma can have antibodies (p.216), known as anti-A (or a) and anti-B (or b). Group A plasma has antibody b; group B has antibody a; there are no antibodies present in group AB, and group O has both a and b antibodies. Agglutination occurs when, for example, plasma with an antibody b mixes with red blood cells with antigen B. In a transfusion, blood can only be used where the plasma of the recipient does not agglutinate the red blood cells of the donor. **compatibility** (*n*).

blood group	
A	antigen A antibody anti-B
B	antigen B antibody anti-A
AB	antigens A and B no antibodies
O	no antigens antibodies A and B

compatible blood groups	recipient blood group			
	O	**A**	**B**	**AB**
O	▽	▽	▽	▽
A	●	▽	●	▽
B	●	●	▽	▽
AB	●	●	●	▽

(left axis label: donor blood group)

KEY

compatible transfusion	▽
incompatible transfusion	●

AB is universal recipient
O is universal donor

incompatible (*adj*) of blood, not suitable for giving in transfusion.

donor (*n*) an individual from whom blood, tissue or an organ is taken to transfer to another.

universal donor an individual with the blood group O who can donate blood to individuals with blood of any group.

recipient (*n*) an individual who receives blood, tissue or an organ from a donor (↑).

universal recipient an individual with the blood group AB who can receive blood from any other group in a transfusion.

respiratory system a system of tubes and spaces in an animal by which air or water is taken in to supply oxygen to the body, and carbon dioxide is expelled out of the body. The exchange of oxygen for carbon dioxide takes place in lungs or tracheae for air, or gills for water.

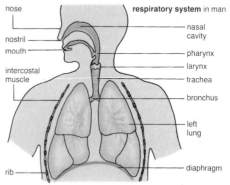

respiratory system in man

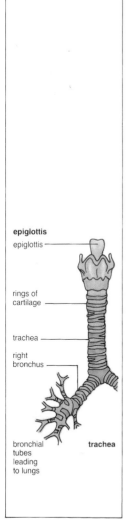

epiglottis

trachea

breath (*n*) (1) one action of taking air into the lungs and expelling the air. (2) the air expelled from the lungs during respiration.

nostril (*n*) an opening in the nose; one of a pair which lead through the nasal cavity to the mouth. Nostrils are also called **nares**.

epiglottis (*n*) a flap of cartilage covered in mucous membrane, above the **glottis**, the opening to the trachea. When swallowing food, the glottis is pushed up against the epiglottis, and the trachea is closed to prevent food entering it.

trachea (*n. pl. tracheae*) (1) in land-living vertebrates, the tube passing from the nose and mouth down the neck to the chest, where it divides into two bronchi (p.176). It carries air to the lungs, and its walls contain C-shaped plates of cartilage which support it. (2) in insects, a tube of the **tracheal system** which carries air from surface openings (spiracles) to the body tissues.

lung (*n*) either of the respiratory organs present in air-breathing terrestrial animals. One is situated either side of the heart. In the lung, oxygen is absorbed from the air by the blood and carbon dioxide is released from the blood.

pleura (*n. pl. pleurae*) one of the membranes which surround each lung in mammals and cover the inside of the thorax. It produces a fluid which lubricates the surfaces in the regions of contact between the lungs and thorax and so prevents friction.

bronchus (*n. pl. bronchi*) one of the two tubes into which the trachea (p.175) divides in air-breathing vertebrates. Each tube leads to a lung and repeatedly splits to form smaller bronchi and finally bronchioles (↓). The bronchi contain plates of cartilage which support the walls and have glands that secrete mucus. The walls are lined with cilia (p.153) which propel any dust trapped in the mucus up to the mouth.

bronchioles (*n.pl.*) small tubular branches of bronchi (↑) which carry air to the alveoli (↓).

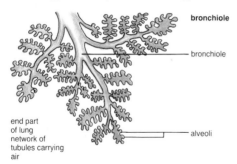

bronchiole

bronchiole

end part
of lung
network of
tubules carrying
air

alveoli

alveolus (*n. pl. alveoli*) one of the numerous minute air-filled sacs in the vertebrate lung. They are situated at the end of each bronchiole (↑) and are covered with a network of blood capillaries. An exchange of oxygen and carbon dioxide takes place through the thin walls.

mucus (*n*) a slimy, viscous secretion produced by mucous membranes which acts as a lubricant.

pulmonary (*adj*) to do with lungs, e.g. the pulmonary vein carries oxygenated blood from the lungs to the heart.

diaphragm (*n*) a dome-shaped sheet of muscle and tendon between the thorax (p.154) and abdomen (p.154) in mammals. Air is drawn into the lungs by the contraction of the muscle which flattens the diaphragm.

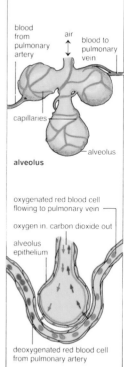

blood
from
pulmonary
artery

air

blood to
pulmonary
vein

capillaries

alveolus

alveolus

oxygenated red blood cell
flowing to pulmonary vein

oxygen in. carbon dioxide out

alveolus
epithelium

deoxygenated red blood cell
from pulmonary artery

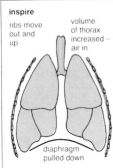

inspire

ribs move out and up | volume of thorax increased – air in

diaphragm pulled down

ribs move in and down | volume of thorax decreased air out

diaphragm relaxed and expire curved up

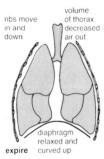

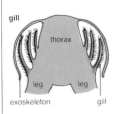

gill

thorax

leg | leg

exoskeleton | gill

cross-section of crayfish in thoracic region

inspire (*v*) to take air into the lungs (p.175) or water into gills. In vertebrates, movements of the ribs draws air into the lungs. In mammals, the diaphragm (↑) also assists to inspire air.

expire (*v*) to expel air from the lungs or water from the gills. In mammals, air is expelled from the lungs by the relaxation of the rib muscles and diaphragm (↑). **expiration** (*n*).

inhale (*v*) (1) to take air into the lungs; the focus is on the purpose of inhaling, e.g. to inhale an antiseptic, or other substance, to cure respiratory disorders. (2) of aquatic animals, to take water into the gills for respiration or other purposes, such as feeding. **inhalation** (*n*).

exhale (*v*) (1) to expel air or other gases from the lungs; the focus is on the purpose of exhaling, e.g. to exhale unpleasant gases. (2) of aquatic animals, to expel water for respiration or other purposes, e.g. an octopus exhales water to propel itself. **exhalation** (*n*).

vital capacity the largest change in volume that can be made by the lungs from filling them completely to emptying them by forced expiration. In man it is about $4000\,cm^3$.

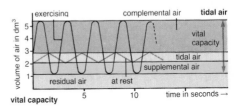

tidal air the volume of air entering and leaving the lungs under normal breathing conditions. In man it is about $500\,cm^3$.

gill (*n*) the respiratory organ of aquatic animals. One gill is usually situated on each side of the animal. Gills have a large surface area and are well supplied with blood. Diffusion of oxygen and carbon dioxide takes place through the thin membranes separating the gill surface and blood capillaries. Gills occur both internally and externally.

branchial (*adj*) concerned with the gills, e.g. branchial arch.

respiration (*n*) the complete breathing process in which oxygen is inspired and carbon dioxide is expired using lungs, gills or tracheae. **respire** (*v*). **respiratory** (*adj*).

external respiration the taking in of oxygen from the environment for use in cells, and the returning of carbon dioxide to the environment.

internal respiration the use of oxygen in cells to provide energy for metabolism.

tissue respiration an alternative term for **internal respiration** (↑)

aerobic respiration takes place in the tissues

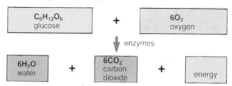

| $C_6H_{12}O_6$ glucose | + | $6O_2$ oxygen |

↓ enzymes

| $6H_2O$ water | + | $6CO_2$ carbon dioxide | + | energy |

aerobic respiration internal respiration (↑) in which oxygen from the air or oxygen dissolved in water is used to oxidize glucose, in a cell, to carbon dioxide and water. Energy is released in the chemical process, and more energy is released by aerobic respiration than by anaerobic respiration (↓).

anaerobic respiration a form of internal respiration in cells, in which no oxygen is used to decompose glucose, or other organic substances. Glucose is decomposed by yeasts to ethanol; in plants this is usually called **fermentation**. In muscles, glucose is decomposed to lactic acid when oxygen is lacking, building up an oxygen debt and producing the sensation of fatigue. In these chemical processes, energy is released, but less energy than is released during aerobic respiration (↑).

anaerobic respiration in muscle cells

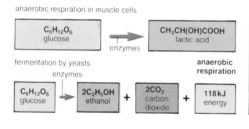

| $C_6H_{12}O_6$ glucose | → | $CH_3CH(OH)COOH$ lactic acid |

enzymes

fermentation by yeasts

anaerobic respiration

enzymes

| $C_6H_{12}O_6$ glucose | → | $2C_2H_5OH$ ethanol | + | $2CO_2$ carbon dioxide | + | 118 kJ energy |

respiration

| respiratory adaptations in the animal kingdom |

| PROTOZOA no respiratory organelles, diffusion of gases |

| COELENTERATES no respiratory organelles, diffusion of gases |

| FLATWORMS no respiratory organelles, diffusion of gases |

| ANNELIDS, e.g. earthworm exchange of gases at cuticle surface which has a rich supply of blood vessels; oxygen carried in circulatory system by pigment – haemoglobin in plasma |

| CRUSTACEA, e.g. crayfish gills; pigment in circulatory system – haemocyanin |

| INSECTS tracheae, sometimes ventilated by general muscle movements of body |

| SPIDERS lung books |

| MOLLUSCS land and freshwater: pulmonary cavity marine: gills and body surface |

| ECHINODERMS diffusion of body surface especially tube feet and gill-like structures |

| VERTEBRATES all have organs and respiratory pigment |

| FISH-gills |

| AMPHIBIANS - lungs, body surface and mouth lining |

| REPTILES - lungs |

| BIRDS - lungs and air sacs |

| MAMMALS - lungs |

start

after several hours fluid has risen due to osmosis

strong sugar solution

water

semipermeable membrane

thistle funnel

experiment to demonstrate
osmosis

turgor
turgid plant cell

cell wall   cytoplasm

plasmolysis
plasmolysed plant cell

cell wall   cytoplasm shrinking from cell wall

homeostasis (*n*) the maintenance of a constant, balanced internal environment. Equilibrium is maintained by the coordinated and automatic response of the organism to any changes, such as a fall in the temperature of its environment. Homeostasis maintains a constant composition of the blood with respect to osmotic pressure, pH values (p.111), and the concentration of various substances, such as glucose and amino acids.

excretion (*n*) the removal of waste products from the body. These are the end-products of nutrition and respiration; they include carbon dioxide from internal respiration (↑), urea from protein nutrition, and water. Excretion takes place from the kidneys, the lungs and the skin.

osmosis (*n*) the process by which molecules of a solvent pass through a semipermeable membrane from a dilute solution to a concentrated solution, but molecules of the solute cannot pass through the membrane. Solvent molecules will pass through the membrane until equilibrium is reached with equal concentrations of solute on each side.

osmotic pressure the pressure exerted by a solvent passing through a semipermeable membrane during osmosis (↑); it is measured by the counterpressure that must be applied to the solution just to prevent the passage of the solvent molecules.

turgor (*n*) the state of a plant cell when fully expanded and unable to absorb any more water. The cell absorbs water by osmosis (↑); as more water enters the cell, the vacuole (p.151) expands, pushing the cytoplasm (p.150) outwards, and exerting turgor pressure on the relatively inelastic cell wall. When turgor pressure equals the osmotic pressure of the cytoplasm, no more water can enter the cell. Turgor pressure keeps the plant cell rigid.

plasmolysis (*n*) the loss of water from a cell by osmosis; it occurs when a plant cell is placed in a solution that is more concentrated than the cell sap. Water is lost from the vacuole (p.151) and the cytoplasm (p.150) is pulled away from the cell wall. There is no outward pressure on the cell wall, and turgor (↑) is lost. **plasmolyze** (*v*).

wilt (*v*) of plants, to lose a greater amount of water than is absorbed, and hence to lose turgor. Excessive water loss will result in the death of the plant.

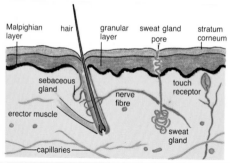

Malpighian layer · hair · granular layer · sweat gland pore · stratum corneum · sebaceous gland · touch receptor · nerve fibre · erector muscle · sweat gland · capillaries

section through human **skin**

skin (*n*) the outer layer covering an animal. It protects the internal tissues (p.150), and is the sense organ for touch, temperature and pressure; it also regulates water loss. In warm-blooded (↓) animals, it helps to control the body temperature.

warm-blooded (*adj*) describes an animal which maintains a constant body temperature, usually higher than that of its surroundings. Birds and mammals are warm-blooded. Sweating provides a means of keeping body temperature constant.

homoiothermic (*adj*) warm-blooded (↑).

sweat gland a gland (p.161) which secretes sweat (↓). In mammals, it consists of a coiled tube surrounded by blood capillaries; sweat is formed by the extraction of water and mineral salts from the blood. A narrow tube, a duct, transports the sweat to an opening at the surface of the skin, through which it is passed. The evaporation of sweat cools an organism; it is a means of controlling the temperature of a warm-blooded (↑) animal.

sweat (*n*) a liquid secreted by sweat glands (↑). It consists of a dilute solution of sodium chloride (common salt), and tiny amounts of other mineral salts.

hair (*n*) a thread-like outgrowth from the skin of a mammal. A hair has a central pith surrounded by a solid cortex, which is enclosed in a thin, hard cuticle. The pith contains air, and at its end, the hair is hollow with no pith. The hair consists of a shaft of dead cells above the skin, and a living, growing root in the skin. The function of hair is to conserve heat.

fur (*n*) a thick covering of soft, fine hair on the skin of some mammals, e.g. cat, lion. Fur provides good protection against heat loss from the body.

hair — hair follicle — nerve fibres — capillary network — root — erector muscle

hair

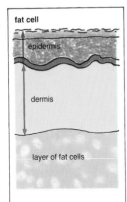

fat cell

epidermis

dermis

layer of fat cells

fat cell a cell containing droplets of oil as a food reserve. A collection of fat cells forms layers of fat in some vertebrates; the fat cells form below the dermis (↓) and provide insulation against heat loss.

cold-blooded (*adj*) describes an animal whose body temperature varies with that of the surrounding air, land or water. Cold-blooded animals cannot control the temperature of their bodies. Except for birds and mammals, all animals are cold-blooded.

poikilothermic (*adj*) cold-blooded (↑).

epidermis (*n*) the outer layer of tissue of a plant or animal. In plants and invertebrates (p.139), the epidermis is only one cell thick. In vertebrates (p.140), the epidermis consists of two layers — the horny layer (↓) and the Malpighian layer (↓).

horny layer the outer layer of the epidermis (↑) of land vertebrates (p.140). It consists of dead cells which are continually rubbed off and replaced by cells from the Malpighian layer (↓). The horny layer is especially thick on the soles of the feet and the palms of the hands; it protects the body against entry of pathogens (p.215), and heat loss.

stratum corneum another name for **horny layer** (↑).

cornified layer another name for **horny layer** (↑).

epithelium (*n*) a tissue which covers exposed surfaces of an organism, forms glands, or lines the tubes and cavities of the body, e.g. the layer of cells lining blood vessels. **epithelial** (*adj*).

Malpighian layer the layer of the epidermis (↑) below the horny layer (↑) and above the dermis (↓); the cells of the Malpighian layer reproduce by cell division, and replace the dead cells of the layer above them.

dermis (*n*) the very thick layer of cells just below the epidermis (↑). It contains blood capillaries for temperature regulation, fat cells for insulation, collagen fibres for elasticity, and also nerve endings, hair follicles, sebaceous glands (↓) and sweat glands (↑).

sebum (*n*) an oily liquid, secreted by the sebaceous glands (↓), which keeps hair and skin waterproof.

sebaceous gland in mammals, a gland in the dermis that secretes sebum (↑), usually into a hair follicle.

pore (*n*) a tiny opening through which substances may be absorbed or secreted, e.g. the pores of plant leaves. **porous** (*adj*).

sebum

secretion enters hair follicle

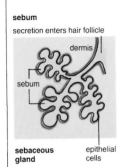

dermis

sebum

sebaceous gland

epithelial cells

kidney (*n*) one of two glands near the spine in vertebrates (p.140). The kidneys control the amount of water in the body, and remove waste substances from the blood which are then excreted as urine (↓) from the bladder (↓).

cortex (*n*) the outer, dark red layer of the kidney (↑); the uriniferous tubules (↓) and Bowman's capsule (↓) are concentrated in it.

medulla[1] (*n*) the inner, paler layer of the kidney, surrounded by the cortex (↑); in it are situated the collecting tubules which drain into the pelvis (↓).

Bowman's capsule a cup-shaped organ in the cortex (↑) of a kidney (↑). Its function is to filter out from the knot of blood capillaries in the capsule such substances as urea (↓), glucose (p.164), mineral salts (p.110) and water. These substances are then led away along a uriniferous tubule (↓).

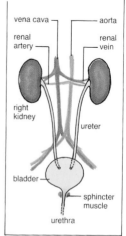

kidney
external
view

urine
out
through
ureter

capsule

cortex

medulla

pelvis

ureter

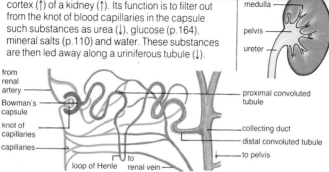

from renal artery

Bowman's capsule

knot of capillaries

capillaries

loop of Henle

to renal vein

proximal convoluted tubule

collecting duct

distal convoluted tubule

to pelvis

uriniferous tubule a narrow, coiled tube in the kidney, leading from the Bowman's capsule (↑) to collecting ducts in the medulla (↑) which carry the urine (↓) to the ureter (↓). As the filtrate from the Bowman's capsule passes along the tubule, blood capillaries which surround the tubule reabsorb substances, such as glucose, mineral salts and water. This reabsorption ensures that only waste substances are excreted in the urine. The amounts of substances absorbed back into the blood vary so that the composition of the blood is kept constant. *See* **homeostasis** (p.179).

pelvis (*n*) a small cavity in the kidney, the upper end of a ureter (↓). It collects urine from all the tubules in the kidney, and delivers the urine to the ureter.

urinary bladder an expandable sac for storing urine (↓). The urine enters from the ureters (↓) and is expelled via a duct called the urethra (↓).

bladder (*n*) the same as **urinary bladder** (↑).

vena cava

aorta

renal artery

renal vein

right kidney

ureter

bladder

sphincter muscle

urethra

sphincter (*n*) a circular muscle which closes the opening to a tube or vessel, e.g. the sphincter closing the exit from the urinary bladder to the urethra. When the sphincter relaxes, the orifice is open.

ureter (*n*) a tube that conducts urine from one of the kidneys to the urinary bladder.

urethra (*n*) in mammals, a narrow tube that carries urine (↓) from the urinary bladder (↑) to the exterior. In males, the urethra is joined by the vas deferens (p.209) and passes through the penis (p.209); it passes both urine and sperm to the exterior.

renal (*adj*) of, or pertaining to, the kidneys.

uriniferous (*adj*) describes a tissue that produces urine (↓).

diuretic (*n*) any substance causing an increase in the production of urine. **diuretic** (*adj*).

ammonia (*n*) a compound with the formula NH_3, formed in the body by the decomposition of amino acids (p.163). Ammonia is poisonous and would cause death if allowed to accumulate in the body. It is made harmless by converting it to urea (↓) in the liver, and then excreting the urea.

deamination (*n*) the removal of an amino group ($-NH_2$) from amino acids (p.163). In mammals, deamination occurs mainly in the liver; the ammonia (↑) formed is converted by the liver into urea (↓).

urea (*n*) a nitrogen-containing water-soluble substance found in urine (↓). It is produced from ammonia (↑) formed in the liver from the decomposition of amino acids (p.163).

uric acid an organic acid, slightly soluble in water. It is formed from ammonia produced in deamination of proteins and nucleic acid (p.210). Birds and insects excrete uric acid for the disposal of nitrogenous waste. Mammals excrete some uric acid in urine.

urine (*n*) a liquid which consists of water, with dissolved organic salts and urea (↑), removed from the blood by the kidneys and excreted through the urethra or cloaca (p.160).

excrete (*v*) to expel the waste products of metabolism from cells or from the body. In simple organisms, the waste products pass out through the cell membrane (p.150). Multicellular organisms get rid of their waste products through the kidneys, lungs or gills, and skin, e.g. urea is excreted by the kidneys.

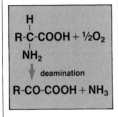

urea

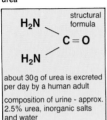

about 30g of urea is excreted per day by a human adult

composition of urine - approx. 2.5% urea, inorganic salts and water

urine

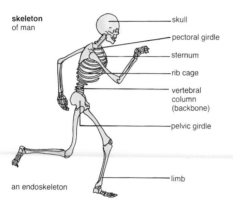

skeleton
of man

skull

pectoral girdle

sternum

rib cage

vertebral
column
(backbone)

pelvic girdle

limb

an endoskeleton

skeleton (*n*) the hard tissue that forms the framework
of an animal body. It supports the body, protects
the internal organs and gives shape and rigidity to
the animal. The skeleton also provides anchorage
for muscles and levers for movement. The two basic
types are exoskeleton (↓) and endoskeleton (↓).

endoskeleton (*n*) a skeleton of bones which is inside
the body of an animal, e.g. the skeleton of a
vertebrate. It provides points of attachment for
muscles, supports internal organs, and provides
protection for most organs. An endoskeleton also
gives shape to an animal's body.

exoskeleton (*n*) a skeleton which is outside the body
of an animal, e.g. the skeleton of an insect, the shell
to molluscs and crustaceans. An exoskeleton can
also be situated in skin, e.g. the bony plates of a
tortoise, the scales of a fish. An exoskeleton
supports and protects internal organs, and
provides points of attachment for muscles.

axial skeleton the part of the endoskeleton (↑)
consisting of the skull (p.186), vertebral column
(p.186), to which is attached the rib cage.

appendicular skeleton the part of the endoskeleton
consisting of the four limbs.

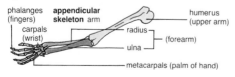

phalanges
(fingers)

**appendicular
skeleton** arm

carpals
(wrist)

radius

ulna

(forearm)

humerus
(upper arm)

metacarpals (palm of hand)

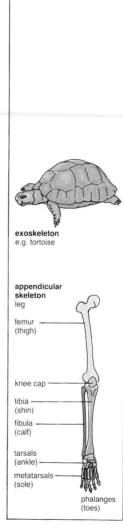

exoskeleton
e.g. tortoise

**appendicular
skeleton**
leg

femur
(thigh)

knee cap

tibia
(shin)

fibula
(calf)

tarsals
(ankle)

metatarsals
(sole)

phalanges
(toes)

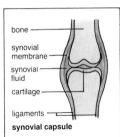

bone

synovial membrane

synovial fluid

cartilage

ligaments

synovial capsule

joint (*n*) a place where two or more bones in the skeleton (↑) are united, usually so that they can move.

synovial capsule a capsule enclosing a freely movable joint (↑). It is attached to the bones on either side of a joint and is filled with synovial fluid which lubricates the joint.

ligament (*n*) a band of strong fibrous tissue between two bones of a joint (↑). It restricts movement of the joint to certain directions and prevents dislocation.

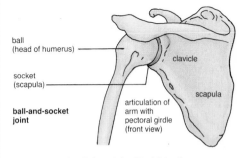

ball (head of humerus)

clavicle

socket (scapula)

scapula

ball-and-socket joint

articulation of arm with pectoral girdle (front view)

ball-and-socket joint a joint (↑) which allows movement in three dimensions, e.g. shoulder joint. The rounded ball-shaped end of one bone fits into a cup-like socket on the other bone.

hinge joint a joint (↑) which allows movement in one plane only, e.g. elbow joint. The round end of one bone articulates with the concave end of the other bone.

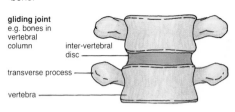

gliding joint e.g. bones in vertebral column

inter-vertebral disc

transverse process

vertebra

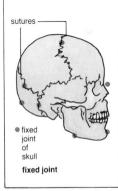

sutures

● fixed joint of skull

fixed joint

gliding joint a joint which allows only limited movement. The surfaces of the two bones slide over each other, e.g. bones in the vertebral column.

fixed joint a joint which does not allow movement, e.g. the joints of bones in the skull (p.186).

skull (n) the bone structure which protects the brain and gives shape to the face.

vertebra (n. pl. vertebrae) a bone consisting of a **centrum**, a central mass of bone, with a **neural arch** enclosing a neural canal. The spinal cord is enclosed, and protected, by the neural arch. Differences occur in the vertebrae of different sections of the vertebral column. **vertebral** (adj).

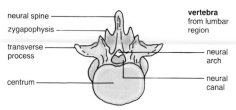

neural spine
zygapophysis
transverse process
centrum

vertebra from lumbar region
neural arch
neural canal

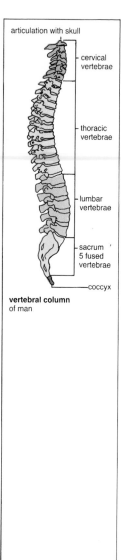

articulation with skull
cervical vertebrae
thoracic vertebrae
lumbar vertebrae
sacrum 5 fused vertebrae
coccyx

vertebral column of man

vertebral column the flexible jointed column which is the chief support for the body of vertebrates. Each bone in the column is a vertebra and is joined to the next by ligaments (p.185) and separated by cartilage (↓). Muscles are attached to the vertebrae. The vertebral column runs from the skull to the tail, and protects and encloses the spinal cord (p.193).

rib (n) a flattened, curved bone (↓) fixed to the vertebral column at one end and either to the sternum (breastbone), another rib, or free at the other end. Ribs partially encircle and protect the thoracic cavity, forming a rib cage.

intercostal muscle a muscle (p.188) situated between two ribs (↑). The ribs are raised by the contractions of these muscles and air is drawn into the lungs (p.175).

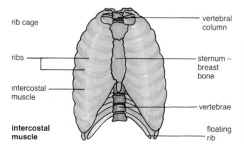

rib cage
ribs
intercostal muscle

intercostal muscle

vertebral column
sternum – breast bone
vertebrae
floating rib

bone

bone (femur)

cartilage

ligament

bone (tibia)

knee joint with knee cap removed

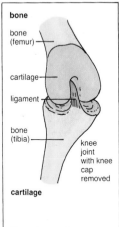

cartilage

Haversian canal

carrying nerves and blood vessels

living bone cells

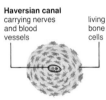

connective tissue

white fibres

ground substance

yellow fibres

macrophage

mast cell

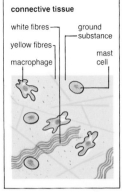

bone (*n*) a connective tissue (p.150) forming the skeleton of vertebrates. It consists of cells embedded in a matrix (↓) of calcium salts and collagen fibres. The fibres give the bone its tissue strength and the calcium salts provide the hardness. The bone cells are connected by fine channels running through the matrix which carry the nerves and blood supply.

matrix (*n*) a non-living, intercellular substance in which living cells are embedded, e.g. bone cells embedded in a bone matrix. The material of the matrix is secreted by the cells in the matrix.

cartilage (*n*) a hard, flexible, connective tissue (↓) with cells distributed in an elastic matrix (↑) formed from protein fibres and containing polysaccharides. The matrix contains no blood vessels. There are several types of cartilage. **Calcified cartilage** is stiffer and forms the entire endoskeleton (p.184) of some animals, e.g. sharks. **Hyaline cartilage** forms the rings in the walls of the trachea (p.175) and bronchi (p.176). **Elastic cartilage** occurs in the outer ear of mammals. **Fibro-cartilage** occurs in the discs between the vertebrae (↑) of mammals.

Haversian canals the fine channels in bone (↑) that carry the nerves and blood vessels. The living cells in bone lie in circles around them.

yellow marrow — long bone

red marrow — **marrow**

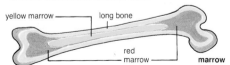

marrow (*n*) the connective tissue (↓) found inside mammalian bones. Yellow marrow is found in the centre of the long bones. It consists mainly of fat cells and produces white blood cells (p.171). Red marrow is present in spongy bone and produces red blood cells (p.170).

ossification (*n*) the formation of bone tissue from the process of changing cartilage. The skeleton of a young child develops from cartilage into bone.

connective tissue a vertebrate tissue consisting of cells, fibres or a matrix (↑). It supports organs and other tissues and has nerves and blood vessels embedded in it. Cartilage (↑) and bone (↑) are connective tissues.

muscle (*n*) a tissue formed from cells which have the ability to contract when stimulated by a motor nerve. All movements of joints are caused by muscle contractions.

fibre (*n*) (1) a strong, thread-like structure formed from protein and found in animal tissues, e.g. nerve fibre. (2) a long, thread-like structure in plants formed from cellulose.

tendon (*n*) a piece of non-elastic connective tissue (p.187) in the form of a band or threads joining a muscle to a bone. It consists of parallel fibres (↑).

voluntary muscle a contractile tissue in vertebrates composed of muscle fibres containing many nuclei. Each fibre is covered with a thin membrane (p.150) and a bundle of muscle fibres are contained within a muscle sheath. Voluntary muscles are controlled by the brain which transmits stimuli via the nerves. They are able to contract very rapidly and powerfully for short periods, after which time they show fatigue. The contractions of the voluntary muscles cause the movement of the skeleton (p.184).

striped muscle an alternative name for **voluntary muscle** (↑), so called because the striations in the cytoplasm (p.150) give the muscle a striped appearance.

involuntary muscle in vertebrates, tissue composed of individual muscle cells with one nucleus and no striations. The cells are spindle-shaped and are bound together with connective tissue (p.187) to form sheets that surround hollow organs, e.g. bladder; alimentary canal. Involuntary muscles are controlled by the autonomic nervous system. They contract at a slower rate than voluntary muscles (↑), but can maintain the contractions for long periods without showing fatigue.

smooth muscle an alternative name for **involuntary muscle** (↑).

cardiac muscle the type of muscle (↑) found only in the wall of the vertebrate heart (p.168). It consists of muscle fibres similar in structure to those of voluntary muscles but without a membrane cover. The fibres form a network and each separate cell has a nucleus. Cardiac muscle is capable of autonomic, rhythmic contractions and is under the control of the pacemaker (p.168).

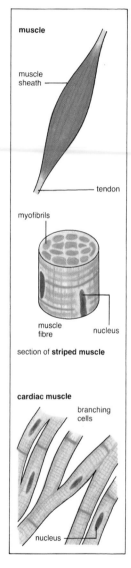

muscle

muscle sheath

tendon

myofibrils

muscle fibre

nucleus

section of **striped muscle**

cardiac muscle

branching cells

nucleus

antagonism (*n*) the action of two muscles producing movement in opposite directions such that the contraction of one muscle relaxes the other. This action is needed to control the movement at a joint, e.g. the biceps muscle contracts and raises the forearm while the triceps relaxes. When the triceps contracts, it straightens the arm as the biceps relax.

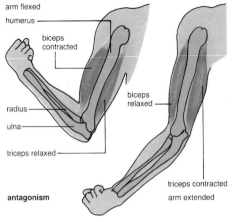

arm flexed

humerus

biceps contracted

biceps relaxed

radius

ulna

triceps relaxed

triceps contracted

arm extended

antagonism

day

leaves in open position

night

leaves in closed position

nastic movement of a plant e.g. leaves of *Oxalis*

movement

movement (*n*) the change of position or posture in a body, e.g. the movement of the neck when the head is turned. In plants, movements are usually growth responses.

locomotion (*n*) the action of moving from place to place, e.g. running. In animals it is brought about by the movement of body parts by muscles (↑).

coordination (*n*) the control of all body functions in animals so they work together for the good of the whole. Body functions of animals are coordinated by the combined action of nerves and hormones.

flex[2] (*v*) to bend; to move the bones of a joint so that the angle between them becomes smaller. The opposite action is to extend a joint.

relax (*v*) of muscles, the opposite of contract. To go into a state of rest from a state of activity. To return to normal length and tension. In antagonistic muscles (↑) one contracts as the other relaxes. **relaxation** (*n*).

nervous system a network of specialized cells in all
animals except unicellular organisms. The simplest
type is a network of cells in small invertebrates,
developing through a nerve cord in larger
invertebrates, to nerve tracts and ganglia, and a
central nervous system (↓). Changes in the external
and internal environment of an animal are detected
by the nervous system. The nervous system
generates impulses in response to stimuli, and
these impulses produce muscular activity or
glandular activity.

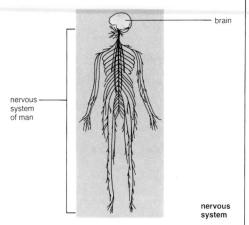

brain

nervous
system
of man

**nervous
system**

central nervous system CNS a mass of nervous
tissue which coordinates all the activities of an
animal. In vertebrates, it consists of the brain (↓) and
spinal cord (p.193). In many invertebrates, a few
large nerve fibres (p.194) joined to several ganglia
serve as the CNS.

brain (*n*) the coordinating centre of an animal's
nervous system. In vertebrates it is protected by the
bones of the skull. In some invertebrates, enlarged
ganglia (p.194) at the anterior end of the nerve cord
act as a primitive brain.

grey matter a type of nervous tissue found in the
central nervous system (↑) of vertebrates. It
contains numerous cell bodies of neurons (p.194),
dendrites, synapses and blood vessels. In the
spinal cord it is internal to white matter (↓), and in

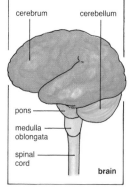

cerebrum   cerebellum

pons

medulla
oblongata

spinal
cord

brain

the cerebellum and cerebral hemispheres it is external to white matter. Coordination in the central nervous system is effected in grey matter.

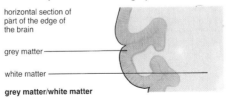

horizontal section of part of the edge of the brain

grey matter

white matter

grey matter/white matter

white matter a type of nervous tissue present in the central nervous system of vertebrates. It consists of nerve fibres (p.194) and connective tissue (p.187). It is external to grey matter (↑) in the spinal cord and internal to grey matter in the cerebellum (p.192) and cerebral hemispheres (p.192).

fore-brain (*n*) the part of the brain (↑) concerned with the senses of smell, vision and maintenance of equilibrium. In higher forms of vertebrates it is concerned with conscious action. In man, it is the largest part of the brain and contains the cerebrum.

mid-brain (*n*) the region of the brain (↑) between the fore-brain (↑) and hind-brain (↓). It is particularly concerned with sight and hearing.

hind-brain (*n*) the region of the brain (↑) that lies between the spinal cord (p.193) and the mid-brain. It is divided into two parts, the medulla[2] (↓) and the cerebellum (p.192).

medulla[2] (*n*) the posterior part of the hind-brain (↑) attached to the spinal cord (p.193). It has grey matter on the inside and white matter on the outside. Its functions include regulating automatic, vital processes such as respiration, heartbeat and dilation of blood vessels. It also controls various reflex actions, such as swallowing and sneezing.

medulla oblongata the full name for **medulla** (↑).

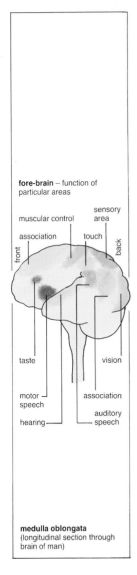

fore-brain – function of particular areas

front

muscular control

association

sensory area

touch

back

taste

vision

motor speech

hearing

association

auditory speech

medulla oblongata
(longitudinal section through brain of man)

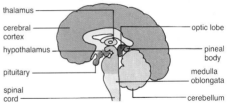

thalamus

cerebral cortex

hypothalamus

pituitary

spinal cord

optic lobe

pineal body

medulla oblongata

cerebellum

autonomic nervous system in vertebrates, a network of nerve-fibres which controls the motor (p.196) functions of the heart (p.168), lungs (p.175), glands (p.161) and other internal organs, and of the involuntary muscles (p.188). It contains two rows of interconnected ganglia (p.194), one on either side of the spinal cord. Its function is to activate important bodily activities automatically.

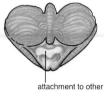

cerebellum

cerebellum viewed from below

attachment to other parts of the brain

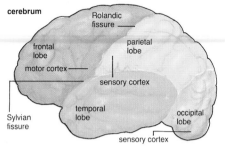

cerebrum

Rolandic fissure

frontal lobe

parietal lobe

motor cortex

sensory cortex

temporal lobe

occipital lobe

Sylvian fissure

sensory cortex

cerebellum (*n*) an outgrowth from the hind-brain (p.191) of a vertebrate. In comparison with other parts of the brain it is small in reptiles and amphibians, larger in mammals, and very large in birds and fish. The grey matter is external to the white matter in mammals. Its function is to maintain balance and coordinate complex muscular movements.

cerebrum (*n*) an outgrowth from the fore-brain (p.191) of a vertebrate. It is the largest brain structure in mammals and is separated into two halves by a fissure. In lower vertebrates it is concerned with smell. In higher vertebrates it controls most of the animal's activities, interprets stimuli from receptors and sends impulses along motor nerves (p.194) causing voluntary actions in effectors. In man, the mental activities of, for example, reasoning and emotion take place in the cerebrum.

cerebral (*adj*) concerned with the brain (p.190) or the cerebrum (↑).

cerebrospinal fluid a clear fluid which fills the cavities in the brain and spinal cord of a vertebrate. It contains glucose and mineral salts, little or no protein, and few cells. Its function is to protect the nervous tissue against mechanical injury and to nourish it.

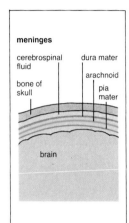

meninges

cerebrospinal fluid

bone of skull

dura mater

arachnoid

pia mater

brain

meninges (*n.pl.*) the collective term for the three membranes that cover the vertebrate brain and spinal cord. They are the dura mater (↓), pia mater (↓), and arachnoid (↓).

dura mater the tough outer membrane covering the brain and spinal cord of vertebrates. It is formed from strong connective tissue (p.187) and contains blood vessels.

pia mater the delicate innermost membrane of the meninges (↑). It contains many blood and lymph vessels (p.167).

arachnoid (*n*) the delicate membrane of the meninges (↑) situated between the dura mater (↑) and the pia mater (↑). It is in contact with the dura mater but separated from the pia mater by spaces filled with cerebrospinal fluid (↑). It contains many blood vessels.

spinal cord a mass of nervous tissue found in vertebrates and containing cell bodies of neurons, nerve-fibres and synapses. It is cylindrical in shape and consists of a core of grey matter (p.190) surrounded by white matter. The spinal cord runs through the vertebrae (p.186) and is protected by them. Paired spinal nerves leave the spinal cord through holes in each vertebra. Simple coordination is effected in the spinal cord by reflex arcs.

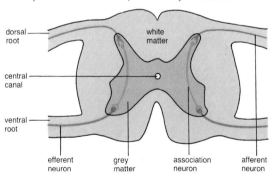

cross-section through **spinal cord**

dorsal root

central canal

ventral root

white matter

efferent neuron

grey matter

association neuron

afferent neuron

peripheral nervous system the nervous tissue not included in the central nervous system (p.190) of an animal. It consists of nerves and nerve-fibres running to and coming from the CNS and branching to every part of the body.

nerve (*n*) a bundle of nerve-fibres and blood vessels enclosed in a sheath of connective tissue (p.187). Each fibre conducts nervous impulses independently. Nerves can vary considerably in length and can be sensory, motor or mixed. **nervous** (*adj*).

nerve-fibre (*n*) an axon (↓) of a neuron (↓) which may be covered with a sheath of myelin (called a medullated nerve-fibre), or not covered (called a non-medullated nerve-fibre). The diameter of a vertebrate nerve-fibre varies from 1 to 20 μm, the length can be as long as the animal, e.g. 2 m or more. In some invertebrates, such as earthworms, the diameter can be up to 1 mm.

neuron (*n*) a cell specialized to conduct nerve impulses in an animal. The body of the neuron cell consists of cytoplasm and contains a nucleus. Thread-like processes (↓) project from the cell body, normally one axon and many dendrons. Some neurons have a single axon and a single dendron. Neurons are in contact with one another through synapses (↓).

ganglion (*n. pl. ganglia*) a solid, swollen mass of nervous tissue on a nerve cord or nerve. It consists of numerous cell bodies of neurons (↑).

process² (*n*) an outgrowth from a cell body, e.g. a dendron.

dendron (*n*) a process (↑), usually short, projecting from the cell body of a neuron (↑). It carries impulses towards the cell body.

dendrite (*n*) one of the fine branches at the terminal end of a dendron (↑). Dendrites receive impulses from axons (↓) of other neurons.

axon (*n*) the long, fine process arising from the cell body of a neuron (↑). It carries nerve impulses away from the cell body. A neuron usually has one axon which terminates at a synapse or an effector, e.g. a muscle.

synapse (*n*) the junction between the end of an axon (↑) of one neuron, and either the dendrites of another neuron or an effector organ, e.g. muscle. One axon can form synapses with several different neurons. A nerve impulse, on reaching a synapse, has to stimulate a connected neuron to continue. Impulses are transmitted through synapses in one direction only, from axon to dendrite.

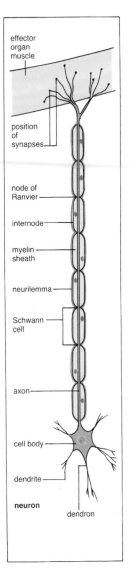

stimulus

pressure
receptor –
Pacinian
corpuscle

the stimulus of pressure
is sensed by Pacinian
corpuscles embedded in
the dermis

stimulus (*n.pl. stimuli*) any change in the internal or external environment of an organism which produces an effect without providing any energy for that effect. In animals, any change in the internal or external environment which provokes an impulse in its nervous system, e.g. touching a hot object gives a sensation of pain. In plants, the presence of water is a stimulus. **stimulate** (*v*).

excite (*v*) to cause, or increase, stimulation of a cell or a tissue. Of a cell or tissue, to respond to a stimulus (↑) or a change in the environment.

inhibit (*v*) to stop or change the rate of any function or action in an animal as a result of control by nerves. **inhibition** (*n*).

irritable (*adj*) refers to an organism which is able to make a response (↓) to a stimulus (↑).

irritability (*n*) the ability of a cell, tissue or organism to respond (↓) to a stimulus (↑). All living organisms have irritability. **irritate** (*v*).

impulse (*n*) a signal, initiated by a stimulus (↑) and conducted along a nerve-fibre. It is like a travelling wave of electrical disturbance and acts along a length of between 2 and 5 cm of nerve-fibre with a speed of between 1 and 100 cm/sec, dependent on the species of animal and the type of nerve. The energy for the impulse is provided by the neuron (↑) so the impulse enters and leaves the neuron with the same amount of energy.

response (*n*) a change in the activity of part, or a whole, of a living organism as a result of a stimulus (↑), e.g. the sight of food stimulates the salivary glands which respond by producing saliva.

reflex (*n*) a simple form of behaviour present in all animals possessing a nervous system. One simple stimulus always causes the same immediate particular response, e.g. the knee jerk.

reflex arc

grey matter   ganglion

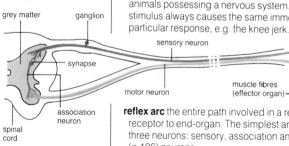

sensory neuron

synapse

stimulus

muscle fibres
(effector organ)

motor neuron

association
neuron

spinal
cord

reflex arc the entire path involved in a reflex (↑) from receptor to end-organ. The simplest arc involves three neurons: sensory, association and motor (p.196) neurons.

automatic (*adj*) refers to a process or action which continues by itself, once started. It is not controlled by external conditions. Respiratory processes and the action of the cardiac (p.188) muscle are automatic.

sensory (*adj*) refers to nerves which conduct impulses, initiated by receptors, to the central nervous system (p.190).

motor² (*adj*) refers to nerves which conduct impulses from the central nervous system to an effector, e.g. muscle, gland. The motor neuron stimulates (p.195) the effector into activity.

association neuron a neuron in the spinal cord which lies between a sensory (↑) and a motor (↑) neuron. Its dendrites are linked by synapses to a sensory neuron axon, and its axon is linked by synapses to dendrites of motor neurons.

sense (*n*) the ability of an organism to receive and react to stimuli (p.195) from its surroundings. Mammals possess five kinds of senses; sight, hearing, taste, smell and touch.

sense-organ (*n*) an organ, containing nervous tissue, specialized for receiving particular stimuli (p.195) and initiating impulses to the brain, e.g. an eye perceives light; nerve endings in the skin detect temperature. A sense-organ responds to only one kind of stimulus, such as light, or sound, or pressure. When a sense-organ receives a stimulus it sends impulses along the nervous system to the brain.

receptor (*n*) an alternative name for a **sense-organ**. A receptor can be taken to mean only the nerve endings which are stimulated, and not the associated tissues of a sense-organ, e.g. the rods and cones of the eye are receptors; the whole eye is the sense-organ.

tactile (*adj*) concerned with, or related to, the sense of touch, e.g. a tactile organ is an organ of touch.

end-organ (*n*) a small organ, composed of one or more cells, located at the end of a nerve-fibre (p.194) which is connected to the peripheral nervous system. It may act as a receptor (↑), or it may change a nervous impulse (p.195) into a stimulus (p.195) for a nerve or gland.

end-bulb (*n*) a small, bulb-shaped organ which is a receptor (↑) in the skin for the sensation of cold.

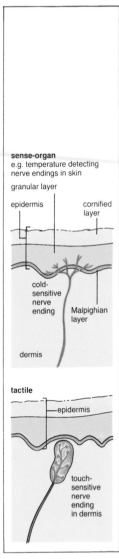

sense-organ
e.g. temperature detecting nerve endings in skin

granular layer

epidermis   cornified layer

cold-sensitive nerve ending   Malpighian layer

dermis

tactile

epidermis

touch-sensitive nerve ending in dermis

pinna (*n.pl. pinnae*) in mammals, the visible part of the ear. It is a flap of skin and cartilage attached to the side of the head, and used to collect sound vibrations and funnel them down into the ear passage. Some mammals can move their pinnae to catch sounds from many different directions, e.g. elephants.

outer ear in mammals, the external part of the ear consisting of the pinna (↑), and the narrow passage leading from the pinna to the ear-drum (↓). Amphibians and some reptiles have no outer ear because the ear-drum is exposed on the skin surface. Birds have no pinnae, and the outer ear consists of a short tube.

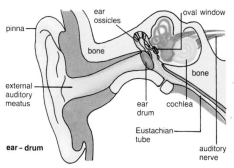

ear-drum

ear-drum (*n*) the thin, tough membrane separating the outer ear (↑) and the middle ear. It vibrates when struck by sound waves, and transmits the vibrations to the tiny ossicles (↓) in the middle ear. The ear-drum covers the external opening of the middle ear in animals, such as amphibians, which do not have an outer ear (↑).

ossicle (*n*) a very small bone in the middle ear (p.198). Mammals, birds, amphibians and reptiles each have a specific number of ossicles in the ear. Mammals possess three ossicles, the malleus, incus and stapes. They transmit the vibrations of the ear-drum (↑) to the oval window (p.198).

Eustachian tube in land vertebrates (p.140), a tube extending from the middle ear (p.198) to the back of the throat. It contains air, and its function is to ensure that the air pressure inside the middle ear is equal to that of the atmosphere.

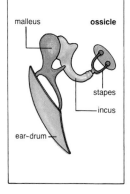

ossicle

oval window a thin membrane between the middle ear (↓) and the inner ear (↓) of mammals, reptiles, amphibians and birds. It transfers vibrations from the ossicles (p.197) of the inner ear to the labyrinth (↓).

middle ear in vertebrates (p.140), other than fish, the air-filled space between the ear-drum and the oval window (↑). It contains the ossicles (p.197).

labyrinth (*n*) part of the inner ear (↓), consisting of a network of tubes and hollows. The membranous labyrinth is filled with liquid and fits inside the bony labyrinth. The labyrinth is concerned with hearing and balance.

semicircular canals in vertebrates (p.140), three fluid-filled curved tubes situated at right angles to each other and connected to the labyrinth (↑). The semicircular canals sense movements of the animal in space and help to keep it balanced.

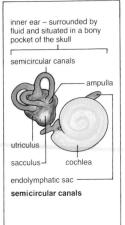

inner ear – surrounded by fluid and situated in a bony pocket of the skull

semicircular canals

ampulla

utriculus

sacculus

cochlea

endolymphatic sac

semicircular canals

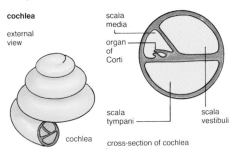

cochlea

external view

cochlea

scala media

organ of Corti

scala tympani

scala vestibuli

cross-section of cochlea

cochlea (*n*) a spiral-shaped tube in the labyrinth (↑). It changes sound vibrations from the liquid in the labyrinth into nerve impulses which are sent to the brain where they are interpreted as sound. The cochlea is able to sense both loudness and pitch.

organ of Corti an organ in the cochlea (↑) which responds to the loudness and pitch of sounds. Hair cells in the organ of Corti respond to vibrations carried in the fluid which enters the cochlea (↑) from the labyrinth (↑). The stimulation of the hair cells initiates impulses which travel along nerve-fibres to the auditory nerve which is connected to the brain, where the pitch of the sound is interpreted.

inner ear in vertebrates, the innermost part of the ear, beyond the middle ear (↑). It contains the labyrinth and semicircular canals.

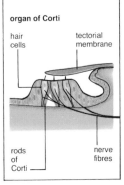

organ of Corti

hair cells

tectorial membrane

rods of Corti

nerve fibres

binocular vision

binocular vision

head of owl viewed from above

monocular vision

eye (*n*) in animals, the sense organ which responds to the stimulus of light. The structure of eyes varies from very simple ones, as found in many invertebrates (p.139), to more complex ones such as the human eye.

eyeball (*n*) in vertebrates, the spherical structure enclosed by the socket and the eyelid. It contains nervous tissue which is stimulated by light.

binocular vision the ability to see an object with both eyes. This allows distance to be judged and shapes to be seen in three dimensions. Monkeys, carniverous vertebrates and humans possess binocular vision.

optic (*adj*) concerned with the eye or the sense of sight, e.g. the optic nerve is a nerve which transmits impulses from the eye to the brain.

tears (*n.pl.*) a salty, sterile, slightly antiseptic liquid secreted by the lachrymal gland (↓). Tears keep the cornea (↓) moist, and wash foreign bodies out of the eyes.

lachrymal gland in land vertebrates, a gland found on the eyeball. It secretes tears. In mammals it is found under the upper eyelid.

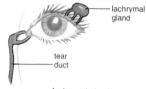

tears flow across eyeball surface and drain into the nasal cavity through the tear ducts

lachrymal gland

tear duct

lachrymal gland/tear duct

longitudinal section through eyeball

conjunctiva

cornea

conjunctiva/ cornea

tear duct a duct, or tube, which drains tears from the eyes into the nose; it is situated at the inner corner of the eye.

conjunctiva (*n*) in vertebrates, a clear, protective membrane covering the cornea (↓) and part of the sclerotic coat (p.200) of the eyeball. It secretes mucus (p.176), and its function is to prevent bacteria entering the eyeball.

cornea (*n*) the transparent covering at the front of the vertebrate eyeball. Towards the outer edge of the eyeball it becomes the sclera. In land vertebrates most of the refraction needed to focus light onto the retina occurs at the cornea.

iris (n) the circular, coloured ring of muscular tissue in the front of mammalian eyes. It has a central opening, the pupil (↓), through which light enters the eye. The iris controls the size of the pupil and hence the amount of light reaching the retina.

pupil (n) the central opening in the iris (↑) through which light passes to the retina. The amount of light reaching the retina controls the size of the pupil by reflex action. The pupil is large in dim light and small in bright light.

crystalline lens the lens of the eye. In vertebrates, it is a transparent structure behind the pupil (↑) separating the aqueous humour (↓) from the vitreous humour (↓). It is attached to a ciliary body which can alter the shape of the lens and so alter the focal length (p.59), helping to focus light onto the retina (↓).

ciliary body the muscular rim at the edge of the choroid coat (↓) which secretes aqueous humour. The iris and suspensory ligaments are attached to it. Ciliary muscle in it can change the shape of the crystalline lens and bring about accomodation.

sclerotic coat, sclera in vertebrates, the tough, opaque, outer layer of the eyeball, made of fibrous or cartilaginous tissue. It helps to protect and maintain the shape of the eyeball. At the front of the eyeball it becomes the transparent cornea.

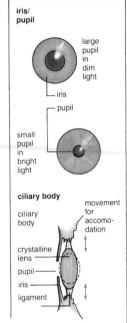

iris/pupil

large pupil in dim light

iris

pupil

small pupil in bright light

ciliary body

movement for accomodation

ciliary body

crystalline lens

pupil

iris

ligament

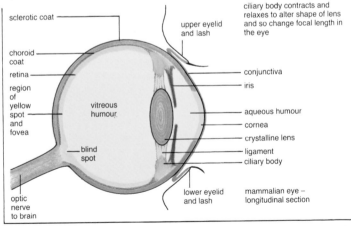

ciliary body contracts and relaxes to alter shape of lens and so change focal length in the eye

sclerotic coat

choroid coat

retina

region of yellow spot and fovea

vitreous humour

blind spot

optic nerve to brain

upper eyelid and lash

conjunctiva

iris

aqueous humour

cornea

crystalline lens

ligament

ciliary body

lower eyelid and lash

mammalian eye – longitudinal section

choroid coat in some vertebrates, the dark-coloured middle layer of the eyeball between the sclera and the retina. It contains blood vessels supplying the retina, and absorbs light, preventing internal reflection.

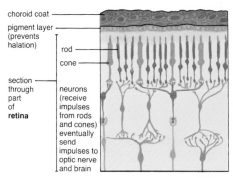

choroid coat

pigment layer (prevents halation)

rod

cone

section through part of **retina**

neurons (receive impulses from rods and cones) eventually send impulses to optic nerve and brain

retina (*n*) in vertebrates (p.140), the light-sensitive layer lining the back of the eye. It is composed of rod cells, cone cells, pigmented cells and nerve cells whose fibres join the optic (p.199) nerve. Light stimulates the rods and cones, and the nerve cells send impulses by the optic nerve to the brain.

fovea (*n*) a small, round depression in the retina of lizards, some birds, apes and humans. Light is mainly focused on the fovea, where vision is acute.

yellow spot an area around the fovea (↑) where the image of an object is focused. It is present in man and in some apes, and is the point at which sight is clearest.

blind spot in vertebrates, a small area at the back of the eye where the optic (p.199) nerve enters the eyeball. Light cannot stimulate nervous impulses at the blind spot because it contains no nerve cells sensitive to light.

aqueous humour a transparent, watery fluid which fills the space in the vertebrate eye between the cornea (p.199) and the crystalline lens (↑). It helps the eyeball to maintain its shape.

vitreous humour a clear, jelly-like material which fills the space between the crystalline lens (↑) and the retina (↑). It gives shape to the eyeball (p.199) and helps the lens by some refraction (p.58) of light.

perception of light by rod cells in **retina**

rod cell containing visual purple (rhodopsin)

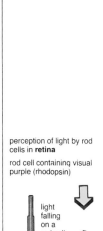

light falling on a rod cell stimulates visual purple to change to another form; an electric impulse is given out at this change

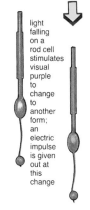

activity of neuron from rod cell

recovery

baseline

impulse

at rest

stimulated by light falling on rod cell

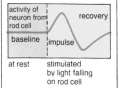

defect (*n*) a missing part, weakness or imperfection that causes a whole to be incomplete, e.g. a defect in an eyeball (p.199) causes imperfect sight.

cataract (*n*) a condition in which the crystalline lens of humans becomes opaque. The lens is removed by a surgical operation and spectacles are used to restore focusing ability.

accomodation (*n*) the action of altering the shape of a crystalline lens, and thus the focusing power of the eye, according to the distance of the object viewed.

accommodation
ciliary body relaxes,
ligament tenses, lens flattens

viewing distant object

viewing near object

ciliary body contracts,
ligament slackens, lens
bulges

uncorrected image falls behind retina

corrected by converging lens in spectacles

light from nearby object

near object

short eyeball

corrected focal length

long sight

long sight a defect (↑) in sight. A person with long sight can see objects clearly at a distance, but near objects look blurred. This occurs because the eyeball (p.199) is too short, and near objects are focused behind the retina (p.201). Long sight is corrected by using converging lenses (p.59) in spectacles.

short sight a defect in sight. A short-sighted person can focus on near objects, but cannot see distant objects clearly. This occurs because the eyeball (p.199) is too long, and distant objects are focused in front of the retina (p.201). Short sight is corrected by using diverging lenses (p.59) in spectacles.

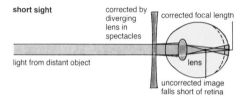

short sight

corrected by diverging lens in spectacles

corrected focal length

light from distant object

lens

uncorrected image falls short of retina

colour blindness the inability to distinguish some or all colours. In humans it is a defect in which certain colours cannot be distinguished. A common type is the inability to distinguish red and green. The defect is inherited and is sex-linked (p.213).

hormone (*n*) (1) in animals, an organic substance produced by endocrine glands (p.204) in very small amounts. Hormones help to coordinate body functions and are carried by the bloodstream to the part of the body on which they act. (2) in plants, an organic substance produced in very small quantities, e.g. an auxin (p.204). The plant cells control the movement of hormones throughout the plant.

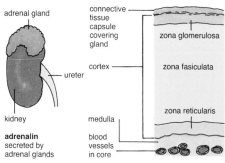

adrenal gland

connective tissue capsule covering gland

zona glomerulosa

cortex

zona fasiculata

ureter

zona reticularis

kidney

medulla

adrenalin secreted by adrenal glands

blood vessels in core

adrenalin (*n*) the main hormone (↑) secreted by the adrenal glands of vertebrates under the conditions of fear, pain, exercise, etc. Adrenalin (a) increases the heartbeat, (b) dilates the blood vessels of the muscles, heart, and brain, (c) contracts the blood vessels of the skin and viscera (p.154), (d) widens the pupil of the eye, (e) increases blood sugar, (f) makes hair stand erect, (g) increases the rate of sweating (p.180). It is antagonistic (p.189) to insulin (↓). Adrenalin is also secreted by some invertebrates (p.139).

insulin (*n*) a hormone (↑) secreted by endocrine cells in the pancreas (p.158) of vertebrates. High concentrations of glucose in the blood stimulate its secretion. Insulin increases the formation of glycogen by the liver and muscles and restricts its conversion to glucose, thus controlling the concentration of glucose in the blood. It is antagonistic (p.189) to adrenalin (↑). A deficiency of insulin causes the disorder **diabetes** in which glucose is found in urine, the kidneys excreting glucose in order to reduce the concentration of glucose in the blood. Diabetes can be treated with injections of insulin.

insulin
secreted by special cells in the pancreas

(acini secrete pancreatic juice)

islets of Langerhans secrete insulin which enters the bloodstream

blood vessel

acini

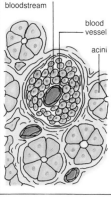

auxin (*n*) any one of a group of plant hormones (p.203) which control growth. Auxins are produced by cells at the growing tips of stems and roots and transferred to the areas of action. Different concentrations cause different rates of growth as in geotropism (p.224), phototropism (p.224), hydrotropism (p.224), etc. Auxins also control growth of fruit and buds, leaf fall and, with other hormones, cell division.

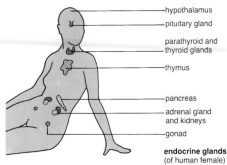

hypothalamus
pituitary gland
parathyroid and thyroid glands
thymus
pancreas
adrenal gland and kidneys
gonad

endocrine glands
(of human female)

endocrine gland in vertebrates, a gland (p.161) that has no duct. It secretes one or more hormones which diffuse into the bloodstream from the gland, e.g. thyroid gland (↓).

ductless gland an alternative name for **endocrine gland** (↑).

thyroid gland in vertebrates, an endocrine gland (↑) which secretes thyroxin, a hormone containing iodine. It is stimulated to produce thyroxin by another hormone secreted by the pituitary gland (↓). Thyroxin regulates the rate of metabolism (p.160) and so controls the growth and temperature of the body. Lack of iodine in the diet causes the disorder **goitre**, in which the thyroid gland becomes enlarged.

pituitary gland in vertebrates, an endocrine gland at the base of the brain, which secretes a number of hormones (p.203). These hormones can be grouped into two main types; those that act directly on parts of the body, and those that stimulate other endocrine glands to produce hormones. The pituitary gland is directly controlled by the central nervous system (p.190) and is the most important endocrine gland.

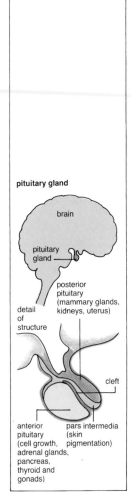

pituitary gland

brain

pituitary gland

posterior pituitary (mammary glands, kidneys, uterus)

detail of structure

cleft

anterior pituitary (cell growth, adrenal glands, pancreas, thyroid and gonads)

pars intermedia (skin pigmentation)

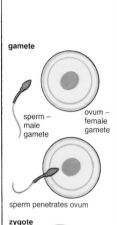

gamete

sperm – male gamete

ovum – female gamete

sperm penetrates ovum

zygote

sperm fertilizes ovum forming a zygote

although several sperm try to penetrate the ovum, only one succeeds; after fertilization the zygote forms a membrane around the outside to prevent other sperm from entering

reproductive system the structures and organs concerned with the production of gametes (↓) and the process of sexual reproduction (↓).

sexual reproduction the fusion (↓) of a male and a female gamete (↓). Usually the male gamete comes from a male member of the species, and the female gamete from a female member. However, many plants and some simple animals, e.g. earthworms, produce male and female gametes within the same organism. Unisexual male and female members of a species have different reproductive systems (↑).

gamete, sex cell (*n*) a male or female reproductive cell with a haploid (p.210) number of chromosomes (p.152) in its nucleus (p.150). The male and female gamete unite to form a zygote (↓) which will develop into a new organism. Usually organisms produce different male and female gametes: the male gamete is generally small and motile, and the female gamete is larger and not capable of locomotion.

gonad (*n*) in animals, the male or female reproductive organ which produces gametes (↑). In some animals the gonads also produce hormones (p.203).

fertilization (*n*) the union of a male and female gamete (↑) during sexual reproduction (↑) to form a zygote (↓) from which a new organism will develop.

external fertilization fertilization (↑) which takes place outside the body of a female. In amphibians and fish, fertilization is external.

internal fertilization fertilization (↑) which takes place inside the body of the female. It occurs in most terrestrial animals.

fusion (*n*) the joining of a nucleus (p.150) of a male gamete (↑) with the nucleus of a female gamete to form a diploid (p.210) nucleus.

zygote (*n*) a cell formed by the fusion (↑) of a male and a female gamete. It is the first cell of a new organism.

fertile (*adj*) of gametes (↑) and organisms, capable of reproducing to form a new organism.

sterile (*adj*) (1) of organisms, not able to reproduce and form a new organism. (2) completely free from bacteria (p.143) and other pathogens.

infertile (*adj*) not able to reproduce under a set of existing conditions.

ovum (*n. pl. ova*) a mature female gamete (p.205) of an animal. Ova are large and non-motile (p.237). They consist of a nucleus (p.150), a large quantity of cytoplasm (p.150), and a yolk (↓) which stores food and water for the developing embryo (↓) after fertilization (p.205).

egg (*n*) (1) an ovum (↑), also known as an **egg cell**. (2) an ovum of any animal, together with a yolk (↓), and enclosed within a protective shell, a tough membrane, or a jelly. Eggs are laid by birds, reptiles, and amphibians.

yolk (*n*) a substance, rich in fat and protein, found in an egg (↑) or yolk sac. The yolk provides food for the developing embryo (↓).

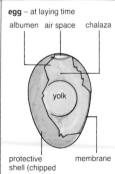

egg – at laying time

albumen   air space   chalaza

yolk

protective shell (chipped away to show contents)

membrane

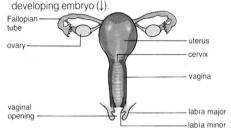

Fallopian tube

ovary

uterus

cervix

vagina

vaginal opening

labia major

labia minor

ovary (*n. pl. ovaries*) a female gonad (p.205) which produces ova (↑). In vertebrates (p.140), it also produces sex hormones (p.203).

Fallopian tube in female mammals, either of the two slender ducts along which ova (↑) travel from the ovaries to the uterus (↓). Tiny hair-like cilia (p.153) waft the ovum along the Fallopian tube.

uterus (*n. pl. uteri*) in female mammals, a hollow, muscular organ in which the foetus (↓) develops and is nourished. The uterus is connected to the vagina (↓) via the cervix (↓). During pregnancy the muscular walls of the uterus increase in thickness; they contract at birth to push the foetus out of the mother's body. In most mammals there are two uteri, one connected to each Fallopian tube (↑). Human beings have only one uterus. **uterine** (*adj*).

womb (*n*) an alternative name for **uterus** (↑).

embryo (*n*) in animals, the earliest stages of the development of a new organism in the uterus. In viviparous animals, the embryo grows within the mother's body. In oviparous animals, the embryo is contained in an egg membrane. **embryonic** (*adj*).

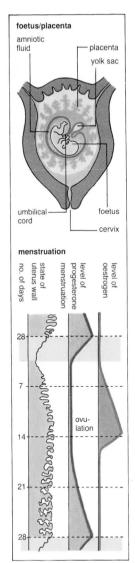

foetus/placenta

amniotic fluid

placenta

yolk sac

umbilical cord

foetus

cervix

menstruation

no. of days

state of uterus wall

menstruation

level of progesterone

level of oestrogen

ovulation

28

7

14

21

28

foetus (*n*) in mammals, the later stages of the development of an organism in the uterus (↑). An embryo (↑) changes to a foetus when it begins to develop recognizable features. In man this occurs after four to six weeks.

umbilical cord a tube connecting the embryo (↑) or foetus (↑) to the placenta (↓) of the mother. The umbilical cord breaks, or is broken, at birth.

placenta (*n*) a special organ in the uterus (↑) of pregnant mammals. The placenta is joined to the foetus by the umbilical cord (↑). In the placenta, the tissues of the foetus and the mother interdigitate (↓), though the blood systems remain entirely separate. It is in the placenta that food substances, oxygen and hormones diffuse from the mother's capillaries (p.170) into the foetal capillaries, and carbon dioxide and waste products diffuse in the opposite direction. At birth the placenta is passed out after the foetus as the afterbirth.

interdigitate (*v*) of two tissues, to grow towards each other; the finger-like projections from one tissue interlock with the projections from the other tissue so that the two structures are linked, but not joined, e.g. the tissues of the mother and foetus interdigitate. **interdigital** (*adj*).

crypts (*n.*) tiny hollows in the walls of the uterus (↑). The placenta (↑) attaches itself to the uterus by growing into the crypts.

menstruation (*n*) in humans, monkeys and apes, the periodic discharge of blood and tissue from the uterus (↑) through the vagina (↓). Menstruation results from the breakdown of the lining of the uterus which occurs when an ovum (↑) has not been fertilized (p.205). In humans, menstruation occurs about every 28 days. After menstruation, the uterine wall begins to thicken and prepares to receive an embryo (↑). **menstruate** (*v*), **menstrual** (*adj*).

cervix (*n*) the neck at the lower end of the uterus (↑) where it joins the vagina (↓). It contains glands which supply the vagina with mucus (p.176). **cervical** (*adj*).

vagina (*n*) in most female mammals, a duct leading from the cervix (↑) to outside the body. It receives sperm (p.209) from the penis (p.209) of the male. At birth, the foetus (↑) is pushed out through the vagina.

hatch (*v*) of offspring (↓), to break out of an egg.
(2) to bring forth young from eggs after incubating
(↓) them.

incubate (*v*) to hatch (↑) eggs by keeping them
warm. **incubation** (*n*).

oviparous (*adj*) describes an animal that lays
fertilized eggs containing an embryo which must
complete its development outside the mother. The
eggs contain a large amount of yolk. Compare
viviparous (↓).

viviparous (*adj*) describes an animal in which the
embryo completes its full development inside the
mother's body. The embryo receives its nutrients
through a placenta (p.207). All placental mammals,
including human, are viviparous.

offspring (*n*) the young produced by living
organisms.

generation (*n*) a single stage in a family history.
Parents form one generation and their offspring the
next generation.

lactation (*n*) the production of milk by the mammary
glands (↓). Hormones (p.203) stimulate the glands
to produce milk. **lactate** (*v*).

mammary glands the milk-producing glands of
female mammals. In most mammals, the glands are
arranged in pairs on the abdomen (p.154). Humans
and apes have only one pair, on the thorax (p.154).

hermaphrodite (*n*) (1) an animal possessing both
male and female gonads (p.205), e.g. earthworm.
(2) a flowering plant possessing male and female
structures on the same flower.

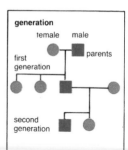

generation

first generation

second generation

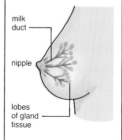

mammary gland

milk duct

nipple

lobes of gland tissue

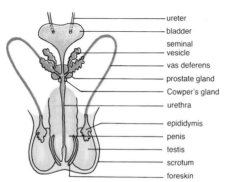

ureter
bladder
seminal vesicle
vas deferens
prostate gland
Cowper's gland
urethra
epididymis
penis
testis
scrotum
foreskin

testis (*n.pl. testes*) a male gonad (p.205) which produces sperm (↓). In vertebrates (p.140) it also produces male hormones (p.203). In mammals, it lies within a pouch of skin that hangs outside the body behind the penis (↓).

testicle (*n*) an alternative name for **testis** (↑).

seminiferous tubules coiled tubes found in the testes of vertebrates. In man, they are about 50cm long and 0.2mm in diameter, and there are several hundred seminiferous tubules contained in a testis. The seminiferous tubules produce sperm (↓).

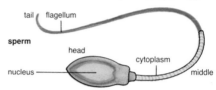

tail / flagellum

sperm

head

cytoplasm

nucleus — middle

sperm (*n*) in animals, a male gamete (p.205) produced in the testes (↑). A sperm cell consists of a nucleus (p.150) surrounded by very little cytoplasm (p.150). In most animals, sperm are able to propel themselves towards the female gamete by means of a flagellum (p.153).

spermatozoon (*n.pl. spermatozoa*) an alternative name for **sperm** (↑).

epididymis (*n*) in vertebrates (p.140), a long coiled tube on the outside of each testis (↑). Sperm (↑) from a testis are stored in the epididymis, and then passed to the vas deferens (↓).

vas deferens (*pl. vasa deferentia*) a muscular duct. In reptiles, birds and mammals it conducts sperm from the epididymis to the cloaca (p.160) or urethra (↓). In fishes and amphibians (p.141), it conducts sperm from a testis (↑) to a cloaca. There is one vas deferens on each side of the body.

urethra *see* **excretory system** (p.179).

penis (*n*) the male sexual organ of all mammals, some reptiles and a few birds. It contains muscles, large numbers of blood vessels, and erectile (↓) tissue which becomes firm so that the male can enter the vagina (p.207) of the female and deposit sperm (↑).

erectile (*adj*) capable of becoming firm and upright when filled with blood.

section through **penis**

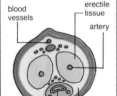

blood vessels

erectile tissue

artery

urethra

muscle and connective tissue

nucleic acid a long chain of many chemically
bonded nucleotides (↓), found in the nucleus
(p.150) of every living cell, which controls the
actions of the cell. There are two types of nucleic
acid, DNA (↓) and RNA (↓).

nucleotide (*n*) an organic compound found in all
living matter. It consists of a sugar (ribose or
deoxyribose), phosphoric acid and a nitrogenous
base. Chains of nucleotides form nucleic acid.

spiral (*n*) a line which curves in a series of concentric
circles of increasing, or decreasing, diameters. The
resulting structure is flat.

helix (*n*) a line which curves in a series of concentric
circles of the same diameter. The resulting structure
is cylindrical.

DNA, deoxyribonucleic acid a type of nucleic acid.
DNA is found in the chromosomes (p.152) of living
cells, and contains the genetic information that
determines the inherited characteristics of an
organism. The nucleotides (↑) which make up DNA
consist of a sugar, deoxyribose; phosphoric acid;
and one of the following nitrogenous bases:
adenine, guanine, cytosine or thymine. The
nucleotides form **strands** which are wound round
each other to form a double helix (↑). The structure
is held together by bonds between pairs of different
bases.

gene (*n*) a tiny segment of a chromosome that
determines a particular characteristic of an
organism, e.g. the colour of hair, the colour of eyes.
Genes are made of DNA (↑). They are present in the
gametes (p.205) and are passed from one
generation to the next, i.e. they are inherited.

diploid (*adj*) describes a nucleus, cell or organism
which has a pair of each type of chromosome
(p.152). Diploid nuclei are found in all cells of an
organism, except gametes (p.205).

haploid (*adj*) describes a nucleus, cell or organism
which contains a single set of chromosomes,
i.e. each chromosome is represented only once.
Haploid nuclei are usually formed as a result of
meiosis (p.152), and are found in gametes (p.205).
The haploid number for a particular organism is half
the diploid number, e.g. humans have a diploid
number of 46 chromosomes per nucleus; the
haploid number, contained in gametes, is 23

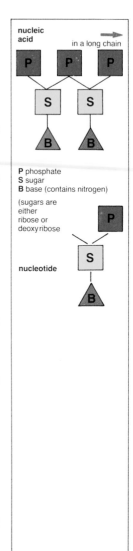

nucleic acid — in a long chain

P phosphate
S sugar
B base (contains nitrogen)

(sugars are either ribose or deoxyribose

nucleotide

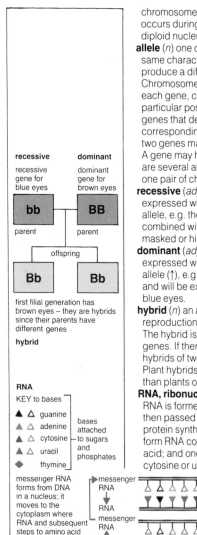

recessive   **dominant**

recessive
gene for
blue eyes

dominant
gene for
brown eyes

bb   **BB**

parent   parent

offspring

Bb   **Bb**

first filial generation has
brown eyes – they are hybrids
since their parents have
different genes

hybrid

RNA

KEY to bases

▲ △ guanine
▲ △ adenine   bases
▲ △ cytosine   attached
▲ △ uracil   and
◆   thymine   phosphates

messenger RNA
forms from DNA
in a nucleus; it
moves to the
cytoplasm where
RNA and subsequent
steps to amino acid
synthesis take place

messenger
RNA

RNA

messenger
RNA

DNA

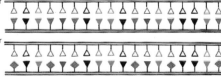

chromosomes. Fusion of two haploid nuclei, as
occurs during sexual reproduction, produces a
diploid nucleus.

allele (*n*) one of a pair of genes (↑) that determine the
same characteristic, e.g. colour of eyes, but
produce a different effect, e.g. brown or green.
Chromosomes are naturally arranged in pairs with
each gene, controlling one characteristic, in a
particular position along the chromosome. Pairs of
genes that determine the same characteristic are in
corresponding positions on each chromosome. The
two genes may be identical, or they may be alleles.
A gene may have more than two alleles, e.g. there
are several alleles determining hair colour, but any
one pair of chromosomes has only two alleles.

recessive (*adj*) describes a gene which will not be
expressed when in the presence of its contrasting
allele, e.g. the gene for blue eyes is recessive when
combined with the gene for brown eyes; it is
masked or hidden by the gene for brown eyes.

dominant (*adj*) describes a gene which will be
expressed when in the presence of its contrasting
allele (↑), e.g. the allele for brown eyes is dominant
and will be expressed when paired with the gene for
blue eyes.

hybrid (*n*) an animal or plant, produced by sexual
reproduction from parents with different genes (↑).
The hybrid is heterozygous (p.212) for one or more
genes. If there are many different genes, as in
hybrids of two different species, the hybrid is sterile.
Plant hybrids are usually larger and more vigorous
than plants of pure strains. **hybridization** (*n*).

RNA, ribonucleic acid a type of nucleic acid (↑).
RNA is formed in the nucleus of cells from DNA and
then passed to the cytoplasm. It is involved in
protein synthesis (p.116). The nucleotides (↑) which
form RNA consist of a sugar, ribose; phosphoric
acid; and one of the four bases, adenine, guanine,
cytosine or uracil.

	generation	genetic composition of parents		phenotype
	A – crossing two homozygous parents			
gametes	first parental	BB ×	bb	brown eyes × blue eyes
		B and B	b and b	
	first filial	Bb	Bb	brown eyes only
	B – crossing two heterozygous parents			
gametes	second parental	Bb ×	Bb	brown eyes × brown eyes
		B and b	B and b	
	second filial	Bb BB bB bb		3 brown eyes, 1 blue eyes

Mendel's laws gametes (p.205) contain only one member of a pair of factors that determine characteristics, such as height and colour. A new organism inherits (↓) one factor from each parent. If one parent is homozygous (↓) for a dominant (p.211) gene, and the other parent is homozygous for a recessive (p.211) gene, their offspring will all have the characteristic determined by the dominant gene, e.g. the offspring of a brown-eyed male BB and a blue-eyed female bb, will all have brown eyes. If the offspring, which are heterozygous for the gene, reproduce, 75% of their offspring will have characteristics determined by the dominant gene, and 25% of the offspring will have the characteristic determined by the recessive gene, e.g. parents who are heterozygous for brown eyes could produce brown- and blue-eyed offspring, the expectation being 3 brown to 1 blue.

heredity (*n*) the passing on of genetic characteristics from one parent to offspring. **hereditary** (*adj*).

pedigree (*n*) a record of inheritance of particular characteristics from one generation to other, later generations.

heterozygous (*adj*) describes a nucleus, cell or organism that possesses two alleles (p.211) for a given characteristic, e.g. a gene for red hair and a gene for brown hair.

homozygous (*adj*) describes a nucleus, cell or organism that possesses two identical genes for a given characteristic.

genotype (*n*) the genetic composition of an organism, i.e. the number and type of genes present, whether expressed or not.

phenotype (*n*) the visible characteristics of an organism.

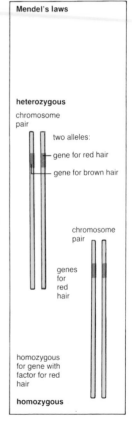

Mendel's laws

heterozygous
chromosome pair

two alleles:
gene for red hair
gene for brown hair

chromosome pair

genes for red hair

homozygous for gene with factor for red hair

homozygous

X-chromosome (*n*) a chromosome (p.152) that determines the sex of an organism. In most animals, including humans, a pair of X-chromosomes will produce a female. Compare **Y-chromosome**.

Y-chromosome (*n*) a sex-determining chromosome. It is shorter than an X-chromosome (↑). A nucleus can contain only one Y-chromosome. In most animals, including humans, an XY-chromosome pair produces a male. Compare **X-chromosome**.

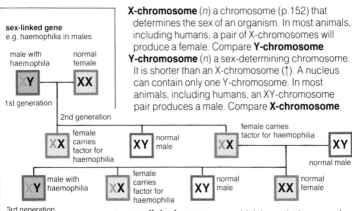

sex-linked gene
e.g. haemophilia in males

male with haemophila

normal female

1st generation

2nd generation

female carries factor for haemophilia

normal male

female carries factor for haemophilia

normal male

3rd generation

male with haemophilia

female carries factor for haemophilia

normal male

normal female

X recessive gene for haemophilia

X dominant normal gene

Y no gene for haemophilia on Y chromosome

i.e. haemophilia is carried by a gene linked to the X chromosome

sex-linked gene a gene which is carried on one of the sex chromosomes, usually the X-chromosome. Sex-linked genes are normally recessive (p.211) and so will not appear if paired with a dominant gene. A male has only one X-chromosome, and is more likely to be affected by a recessive gene carried on an X-chromosome than a female, who has two X-chromosomes. The sex-linked gene cannot be passed on by a male to its male offspring, but a female offspring can inherit (↓) the gene on the X-chromosome, and can pass it on to her son. Thus, a sex-linked gene can appear in male offspring every other generation.

inherit (*v*) to receive genes from parents as a result of reproduction. The information concerning genetic characteristics is contained in the chromosomes in the nucleus (p.150) of gametes (p.205). Offspring are derived from the gametes of their parents. In sexual reproduction the chromosomes in the male gamete pair with those of the female gamete to form a new organism. The offspring thus inherits genes from both parents and its characteristics are determined by both parents.

pass on (*v*) of parents, to give genes to their offspring as a result of reproduction.

clone (*n*) a group of individuals propagated by asexual methods from one sexual ancestor, thus all members being genetically identical. For plants, vegetative propagation is the asexual method.

evolution (*n*) the way in which oganisms have gradually undergone a series of changes over millions of years to produce new species of organisms, usually more complex in form.

mutation (*n*) a sudden variation in the genes (p.210) or chromosomes (p.152) of an organism which usually produces an observable change. Mutations arise from a change in the DNA (p.210) of the chromosomes. This may occur spontaneously, or be caused by radiation, neutrons and certain chemicals. Mutation in the genes of body cells affects only the organism concerned; but mutations in the genes of gametes (p.205) affect all the offspring (p.208). Some mutations benefit the organism, but most have an adverse effect.

mutant (*n*) an organism that contains a gene that has been changed by mutation (↑).

natural selection a theory of evolution (↑). It is based on the observation that the characteristics of individual organisms within a population vary, and there is a tendency for those organisms which are best suited to the conditions under which they live to survive and produce the next generation. A process of selection occurs naturally; individuals with the best adaptive characteristics have more offspring and pass on their characters, whereas those individuals less well adapted tend to have fewer offspring or die out. The population gradually changes as a result of natural selection. The fittest survive, and unsuitable mutants die.

Darwinism (*n*) a theory that evolution (↑) takes place by natural selection (↑).

population (*n*) the total number of organisms of the same type, usually of the same species, that live in a given area at any one time. **populate** (*v*).

survival (*n*) the act, state, or fact of living or existing.

reversion (*n*) of a plant or animal, returning to the characteristics of an earlier ancestor, by a gradual process, e.g. cultivated roses, over successive generations, can revert to wild roses. **revert** (*v*).

extinct (*adj*) describes species of organisms which no longer exist on Earth, e.g. dinosaurs.

fossil (*n*) the hardened remains, or shape, of a plant or animal preserved in rock formations in the Earth's crust. By studying fossils we can learn about organisms that lived millions of years ago.

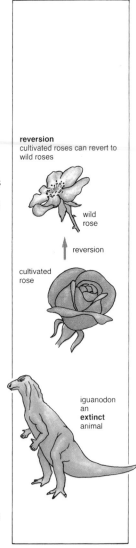

reversion
cultivated roses can revert to wild roses

wild rose

reversion

cultivated rose

iguanodon an **extinct** animal

vitamin deficiency diseases	
vitamin	disease
A	night blindness and xerophthalmia
B_1	beri-beri
B_2	conjunctivitis
B_6	part of protein synthesis cycle lost
B_7	pellagra
B_{12}	pernicious anaemia
C	scurvy and poor wound healing
D	rickets and osteomalacia
E	infertility
K	ineffective blood clotting

deficiency disease

threadworm
gut parasite

pathogens

roundworm
gut parasite

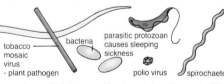

tobacco mosaic virus
- plant pathogen

bacteria

parasitic protozoan causes sleeping sickness

polio virus

spirochaete

disease (*n*) a state of a tissue, organ, system, or organism, during which its function is not carried out normally. Disease is caused by pathogens (↓), infestations, deficiencies in diet, or internal malfunction of tissues or organs, e.g. influenza caused by a pathogen; tapeworm infestations; rickets caused by vitamin deficiency in diet; diabetes caused by malfunction of the pancreas. **diseased** (*adj*).

contagious (*adj*) describes a disease that can be transmitted by personal contact. **contagious** (*n*).

infect (*v*) of pathogens (↓), to enter a tissue, organ or organism and cause disease (↑).

isolate (*v*) to separate a person or animal suffering from a contagious (↑) disease from other people or animals to prevent the disease spreading.

malnutrition (*n*) a condition of undernourishment resulting from a lack of food, a poorly balanced diet or disease (↑).

deficiency disease a disease (↑) caused by a lack of vitamins, minerals or amino acids needed in the diet.

pathogen (*n*) a microorganism that causes a disease (↑), e.g. parasites (p.240) such as bacteria, viruses, protozoa or worms. A pathogen may cause disease in one type of organism, but not another. **pathogenic** (*adj*).

septicaemia (*n*) a condition in which pathogenic (↑) bacteria enter and infect (↑) the blood; the bacteria may come from a wound or an infected tissue.

infest (*v*) to live as parasites (p.240) in large numbers on animals or plants, or on clothes and in buildings.

vector² (*n*) an animal or physical agent which carries a pathogen (↑) from an infected plant or animal to a host (p.240). Physical vectors are water, food, air and physical contact. Animal vectors include mosquitoes, fleas and dogs.

causative agent the pathogen (↑) which causes a disease (↑), e.g. *Tinea*, a type of fungus (p.142) is the causative agent of athlete's foot.

antigen (*n*) a foreign substance, usually a carbohydrate or a protein; it causes an organism to produce a specific antibody (↓), which destroys the antigen and eliminates it from the body.

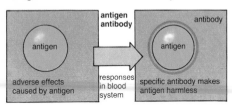

antibody (*n*) a chemical substance made in the blood, usually in response to contact with an antigen (↑). The antibody combines with the antigen and renders it harmless. The production of an antibody is stimulated by the presence of an antigen, and the antibody can generally only combine with one particular antigen.

toxin (*n*) a poison, produced in a plant or animal by microorganisms, which causes certain diseases. **toxic** (*adj*).

antitoxin (*n*) an antibody (↑) which combines with toxins (↑) of disease-producing bacteria and renders the toxins harmless.

penicillin (*n*) a powerful antibiotic produced by certain moulds (p.144), and used to prevent the reproduction of bacteria.

inoculation (*n*) the act of introducing a vaccine (↓) or an antiserum (↓) into the blood of a vertebrate (p.140) in order to make it immune (↓) to a particular disease. **inoculate** (*v*).

vaccine (*n*) a liquid containing dead, inactive or harmless pathogens (p.215). When introduced into the body of a vertebrate (p.140) the vaccine stimulates the production of antibodies. The antibodies help the body to resist future attacks from active pathogens of the same kind.

antiserum (*n*) a serum (p.172) which contains antibodies (↑) to a specific antigen (↑) or pathogen (p.215). The antiserum is taken from the blood of an animal which has produced antibodies in reaction to the antigen. A vertebrate (p.140) inoculated (↑) with the antiserum will have immediate protection against the pathogen.

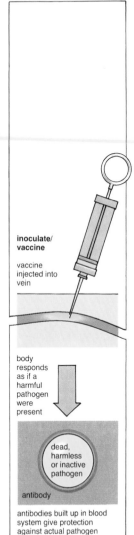

inoculate/ vaccine

vaccine injected into vein

body responds as if a harmful pathogen were present

dead, harmless or inactive pathogen

antibody

antibodies built up in blood system give protection against actual pathogen

immunity (*n*) the ability of a plant or animal to recognize and destroy antigens (↑) or pathogens (p.215). Immunity may be beneficial to an organism, e.g. resistance to, and recovery from, a harmful disease like malaria; or it may harm the organism, e.g. the rejection of transplants and blood transfusions. Organisms provide immunity by (1) not allowing the pathogens to enter through the skin, (2) destroying them in the body by means of phagocytes (p.171), (3) destroying them by the action of acid gastric juices, or (4) by producing antibodies to make antigens harmless. Immunity is needed against each individual disease, or each particular antigen.

immunity

type of pathogen given:
- living attenuated virus
- living attenuated vaccina virus
- virus killed by formalin
- killed bacillus
- living attenuated bacillus
- toxoid

disease	period of immunity →					
tetanus						6 years
cholera	6 months					
influenza	6 months					
tuberculosis				4 years		
smallpox			3 years			
yellow fever						10 years

acquired immunity a type of immunity (↑) acquired by having a disease and producing antibodies (↑) against the pathogens (p.215) or toxins (↑) which caused the disease. The presence of antibodies means that the body is able to resist further attacks of the disease. The length of time an organism will remain immune to a disease varies with the disease, e.g. immunity to a common cold lasts for a short period only; immunity to chicken pox lasts for several years.

artificial immunity a type of immunity (↑) acquired by an inoculation (↑), and usually only lasting for a short time.

passive artificial immunity a type of immunity (↑) acquired by injecting an organism with an antiserum (↑) from another individual. The antiserum already contains antibodies (↑) to a particular disease, so the recipient does not actively produce antibodies. Compare **active artificial immunity** (↓).

active artificial immunity immunity (↑) acquired by injecting an organism with a vaccine (↑). The vaccine stimulates the organism to produce the appropriate antibodies (↑). Compare **passive artificial immunity** (↑).

allergy (*n*) an abnormal reaction of the body to a particular antigen (p.216). Certain foods, pollen, dust, animal furs or drugs can cause allergies. **allergic** (*adj*).

poison (*n*) any chemical substance which can cause death or serious damage to an animal, or a plant, e.g. compounds of arsenic and phosphorous. **poison** (*v*), **poisonous** (*adj*).

venom (*n*) a poison (↑) produced by an animal and introduced into the body of its victim by a bite or sting.

antidote (*n*) a chemical substance that neutralizes the effects of a poison or a venom.

epidemic (*adj*) describes an infectious disease (p.215) which spreads rapidly and affects a large number of people in one region at the same time.

endemic (*adj*) describes a disease (p.215) which is always present in a country or region, or amongst particular groups of people.

pandemic (*adj*) describes an epidemic (↑) which affects a very large area, e.g. a continent.

sporadic (*adj*) describes a disease (p.215) which occurs irregularly in particular areas.

anaemia (*n*) a disease in which there are not enough red cells in the blood, or the cells lack haemoglobin. The blood carries too little oxygen to the tissues, so a person is pale and lacking in energy. **anaemic** (*adj*).

atrophy (*n*) the wasting away of a tissue, organ or part of the body; often caused by lack of use. **atrophy** (*v*).

anaesthetic (*n*) a chemical substance which produces loss of feeling in a part of the body, a **local anaesthetic**, or causes a person to be unconscious, a **general anaesthetic**. Anaesthetics are used in surgical operations and in dentistry. **anaesthetize** (*v*).

fracture (*n*) a crack or break in a bone; there are several types. **Green-stick** in which the bone is only partially broken. **Simple** in which the bone is broken, but damage to surrounding tissues is slight. **Compound** which has a wound going down to the bone, and the bone may protrude through the skin. **Complicated** in which blood vessels or internal organs may be involved. **Impacted** in which the ends of a broken bone are driven into each other.

warning label for containers of **poison**

atrophy
some causes of atrophy

1. lack of use
2. defective supply of nutrients
3. interference with nerve supply
4. deficiency of secretions from endocrine glands
5. action of toxins

air pollutants:
smoke, sulphur dioxide,
fumes and dust from
burning oil, petrol and coal

lead from petrol combustion

water pollutants:
industrial waste, sewage,
fertilizers, pesticides and
detergents

temperature – hot cooling-
water from power stations

oil pollutants:
accidental and deliberate
spillage of heavy oils at sea

radioactive waste

birth rate the number of births per year per thousand
of the population in a given area or group.
death rate the number of deaths per year per
thousand of the population in a given area or group.
mortality rate the number of deaths in a given period
per thousand of the population in a given age group.

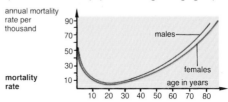

annual mortality
rate per
thousand

mortality
rate

hygiene (*n*) the science of maintaining health and
preventing disease by eating a balanced diet,
taking adequate exercise, ensuring cleanliness,
and avoiding contaminated (↓) food or water.
contaminate (*v*) to spread pathogens, bacteria or
viruses by contact. Water, air, food, clothes and
buildings can be contaminated.
pollute (*v*) to make the atmosphere or environment
unclean and unhealthy, and so harm living
organisms, e.g. car exhaust fumes pollute the air;
factory chemicals can pollute rivers. **pollution** (*n*).
antiseptic (*n*) a chemical substance used on cuts
and wounds to prevent pathogens (p.215) entering
or multiplying. **antiseptic** (*adj*).
aseptic (*adj*) describes a condition in which there
are no pathogens (p.215) present.
pasteurize (*v*) to destroy disease-producing
bacteria in a food or liquid by heating it to a given
temperature for a specified period of time.
pasteurization (*n*).

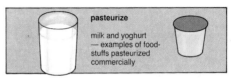

pasteurize

milk and yoghurt
— examples of food-
stuffs pasteurized
commercially

irradiate (*v*) (1) to direct rays of ultraviolet light onto
food in order to kill bacteria, a similar effect to
pasteurization (↑). (2) to direct any form of radiation
onto a substance or an object. **irradiation** (*n*).

plant (*n*) a living organism belonging to the plant kingdom. All plants, except fungi, make their own food by photosynthesis (p.223). Plants respire, grow and reproduce, but, unlike animals, they cannot move from place to place. They react very slowly to stimuli because they have no nervous system.

root (*n*) the underground part of a plant that holds it in position, and absorbs water and salts from the soil. Roots differ from stems (↓) in internal structure, and in not bearing leaves or buds.

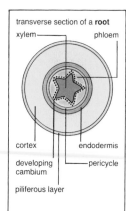

transverse section of a **root**

xylem — phloem
cortex — endodermis
developing cambium — pericycle
piliferous layer

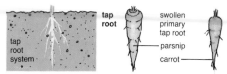

tap root
tap root system
swollen primary tap root
parsnip
carrot

tap root the main root (↑) of a plant; it grows straight downwards and usually has other roots branching out from it. Tap roots develop from the radicle of the seed. Some tap roots store food, e.g. parsnips.

adventitious root a root which develops from any node (↓) on a stem, e.g. in a strawberry. There are several types of adventitious root.

foliage
rhizome (horizontal underground stem) bearing **adventitious roots**
bud
adventitious roots

prop root an adventitious root (↑) which grows from a node (↓) on a stem (↓); it grows downwards into the earth and helps to support the stem.

fibrous root one of many adventitious roots (↑) that grow from the base of a stem (↓), usually with no tap root (↑), in the system of roots.

root hair a fine, hair-like projection from cells on the surface of young roots. Root hairs increase the amount of surface area available for absorbing water and inorganic salts from the earth. As the root grows downwards, the root hairs die.

piliferous layer the layer of cells which bear root hairs; it covers the surface of the root (↑) near the tip.

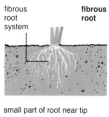

fibrous root system
fibrous root

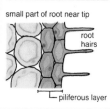

small part of root near tip

root hairs
piliferous layer

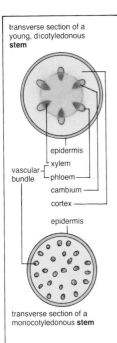

transverse section of a young, dicotyledonous **stem**

epidermis

vascular bundle
— xylem
— phloem

cambium

cortex

epidermis

transverse section of a monocotyledonous **stem**

wood

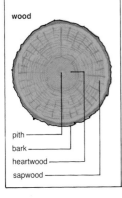

pith

bark

heartwood

sapwood

stem (*n*) the main upward-growing part of a plant that bears leaves, buds (p.225) and flowers. It is usually above ground. Its main function is to provide support, transport food and water, and store food.

stalk (*n*) a long, thin structure, without branches, which supports a leaf or flower.

node² (*n*) the part of a plant stem (↑) from which leaves or aerial roots develop.

internode (*n*) the part of a plant stem (↑) between two successive nodes; no leaves grow on it.

lenticel (*n*) a small, elliptical, raised hole, formed in woody stems, which allows the passage of gases into and out of the stem.

vascular bundle tissue, consisting mainly of xylem and phloem, which conducts water and salts up from the roots, and soluble food substances down from the leaves. Vascular bundles run through stems, leaves and roots. Monocotyledons and dicotyledons differ in the arrangement of their vascular bundles.

xylem (*n*) a woody tissue in the vascular bundle (↑) of a plant; it is made of tiny tubes and carries water and salts from the roots up to the stems and leaves by means of root pressure (p.223) and transpiration (p.223). It also provides support for the plant stem.

phloem (*n*) a tissue in the vascular bundle (↑) of a plant; it contains many tiny tubes which carry soluble (p.47) food substances down the stem from the leaves to other parts of the plant.

cambium (*n*) the cellular layer between the xylem and the phloem of dicotyledons. As the plant grows the cells divide, forming xylem on one side and phloem on the other. As a result, the stem thickens.

cortex (*n*) an outer layer of tissue enclosing the vascular system of dicotyledons. It stores food.

pith (*n*) the soft, spongy core at or near the centre of certain plant stems; it stores food and water.

wood (*n*) the hard, fibrous part of tree and shrub stems; it is found between the bark and the soft inner pith (↑). It is formed from cells whose cellulose walls have been strengthened by a deposit of **lignin**. The cells lose all the cytoplasm as they age; their only function becomes to provide support. **Heartwood** consists of old cells possessing no cytoplasm and unable to conduct water. **Sapwood** is not as strong as heartwood, and conducts some water. **woody** (*adj*).

leaf (*n*) a flat, thin structure, usually green, growing from a node on a stem or twig of a plant. It consists of a stalk and a broad, flattened portion, the blade. Leaves carry out the important functions of photosynthesis (↓) and transpiration (↓).

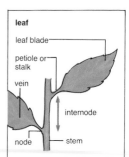

leaf

leaf blade

petiole or stalk

vein

internode

node

stem

margin

laurel

margin

birch

oak

holly

margin (*n*) the border, or edge, of a leaf. Different species of plant have characteristic leaf margin shapes, e.g. a holly leaf; an oak tree leaf.

petiole (*n*) a stalk extending from a node to the base of a leaf.

venation (*n*) the system of tubes, the veins, visible on the surface of a leaf. The veins carry water and nutrients for the leaf and give it support.

foliage (*n*) a general term for all the leaves on a plant, apart from the special leaves, such as cotyledons (p.230) and similar leaf-like parts of plants.

axil (*n*) in plants, the angle between the upper side of the petiole (↑) and the stem from which it arises.
axillary (*adj*).

modified leaf a leaf which has a different size or shape from a common leaf, and also a different function, e.g. the leaf tendrils of grapevines. They have a long, thin shape and support the plant by twisting around objects.

deciduous (*adj*) describes trees and shrubs which shed their leaves every year, e.g. elm trees are deciduous.

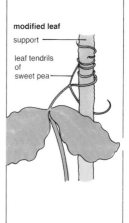

modified leaf

support

leaf tendrils of sweet pea

deciduous

evergreen

evergreen (*adj*) describes trees and shrubs which bear leaves all year round, e.g. pine trees are evergreen.

chlorophyll (*n*) a substance responsible for the green colour in leaves. Chlorophyll, contained in plastids (p.151), takes in the Sun's energy and this is used by the plant to make food by photosynthesis.

photosynthesis (*n*) in green plants, the synthesis of soluble plant foods (organic compounds) from carbon dioxide and water, using energy from sunlight; chlorophyll (↑) absorbs the sunlight and supplies energy for the process. Oxygen is a by-product of photosynthesis.

stoma (*n. pl. stomata*) a tiny pore found in the outer cell layer of certain plant stems and particularly the underside of leaves. Each stoma has two guard cells (↓) round it which control the size of the pore. Plants take in oxygen and carbon dioxide from the air through the stomata, and pass out oxygen, carbon dioxide and water vapour. When the plant is short of water the stomata close.

stoma/guard cell
1. stoma open
2. stoma closed

guard cell ——

leafy shoot
reservoir
capillary tube with bubble of air
rubber bung
water
simple **potometer**

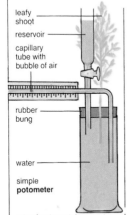

rate of water uptake or water transpired from the shoot, is measured by noting the distance travelled by an air bubble in the capillary tube, at given intervals of time; reservoir allows bubble to be re-started

guard cell one of a pair of crescent-shaped cells round a stoma (↑) which regulates the size of the stoma. As the plant loses water, the turgor pressure is low, and the guard cells contract and close the stoma, preventing further water vapour loss. When turgor pressure is high, the plant has sufficient water, the guard cells expand and become curved, opening the stoma and allowing water vapour to leave the plant.

root pressure the pressure under which water passes from the root cells to the xylem (p.221) vessels of a plant.

transpiration (*n*) the loss of water vapour from the surface of plant leaves, mainly through the stomata. Transpiration helps root pressure (↑) by drawing water up the stem. Excess transpiration is harmful as it causes the plant to wilt. **transpire** (*v*).

sap (*n*) the watery liquid that circulates in plants; it transports food and water to the living cells and carries away waste.

potometer (*n*) an apparatus for measuring the rate at which water is taken up into a shoot, or the rate of transpiration (↑).

tropism (*n*) the tendency of a plant to grow in a particular direction, or to change direction, in response to the influence of external conditions, e.g. light; gravity; water.

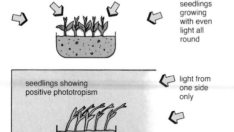

seedlings growing with even light all round

seedlings showing positive phototropism

light from one side only

phototropism

phototropism (*n*) a tropism (↑) of an organism in response to light. It may be a **positive** tropism, as when plant stems grow towards the light; or a **negative** tropism, as with roots which grow away from the direction of light. The phototropism of leaves is always positive. **phototropic** (*adj*).

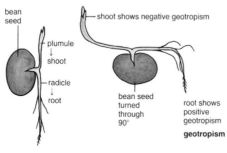

bean seed

plumule ↓ shoot

radicle ↓ root

shoot shows negative geotropism

bean seed turned through 90°

root shows positive geotropism

geotropism

geotropism (*n*) a tropism (↑) of an organism in response to gravity. Negative geotropism refers to the upward growth of a plant stem, and positive geotropism to the downwards growth of the main roots. **geotropic** (*adj*).

hydrotropism (*n*) a tropism (↑) of an organism in response to water. Roots tend to show positive hydrotropism by growing towards water in the soil. **hydrotropic** (*adj*).

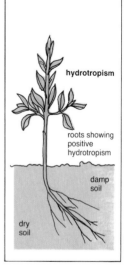

hydrotropism

roots showing positive hydrotropism

damp soil

dry soil

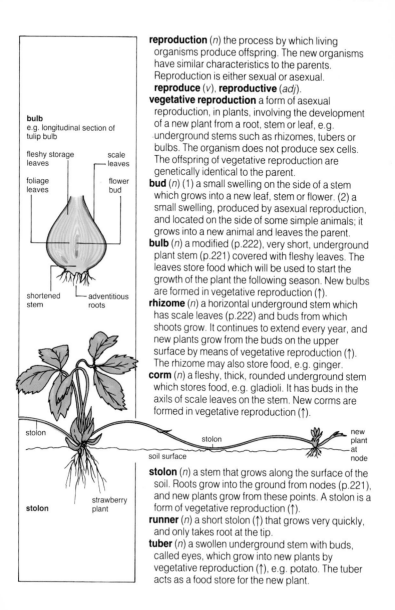

bulb
e.g. longitudinal section of
tulip bulb

fleshy storage
leaves

scale
leaves

foliage
leaves

flower
bud

shortened
stem

adventitious
roots

stolon

stolon

soil surface

new
plant
at
node

strawberry
plant

stolon

reproduction (*n*) the process by which living
organisms produce offspring. The new organisms
have similar characteristics to the parents.
Reproduction is either sexual or asexual.
reproduce (*v*), **reproductive** (*adj*).

vegetative reproduction a form of asexual
reproduction, in plants, involving the development
of a new plant from a root, stem or leaf, e.g.
underground stems such as rhizomes, tubers or
bulbs. The organism does not produce sex cells.
The offspring of vegetative reproduction are
genetically identical to the parent.

bud (*n*) (1) a small swelling on the side of a stem
which grows into a new leaf, stem or flower. (2) a
small swelling, produced by asexual reproduction,
and located on the side of some simple animals; it
grows into a new animal and leaves the parent.

bulb (*n*) a modified (p.222), very short, underground
plant stem (p.221) covered with fleshy leaves. The
leaves store food which will be used to start the
growth of the plant the following season. New bulbs
are formed in vegetative reproduction (↑).

rhizome (*n*) a horizontal underground stem which
has scale leaves (p.222) and buds from which
shoots grow. It continues to extend every year, and
new plants grow from the buds on the upper
surface by means of vegetative reproduction (↑).
The rhizome may also store food, e.g. ginger.

corm (*n*) a fleshy, thick, rounded underground stem
which stores food, e.g. gladioli. It has buds in the
axils of scale leaves on the stem. New corms are
formed in vegetative reproduction (↑).

stolon (*n*) a stem that grows along the surface of the
soil. Roots grow into the ground from nodes (p.221),
and new plants grow from these points. A stolon is a
form of vegetative reproduction (↑).

runner (*n*) a short stolon (↑) that grows very quickly,
and only takes root at the tip.

tuber (*n*) a swollen underground stem with buds,
called eyes, which grow into new plants by
vegetative reproduction (↑), e.g. potato. The tuber
acts as a food store for the new plant.

vegetative propagation a method of producing new plants from an existing plant by vegetative reproduction (p.225). The process uses parts of the plant which are capable of growing by vegetative reproduction, i.e. a root, stem or leaf. *See* **cutting** (↓), **graft** (↓).

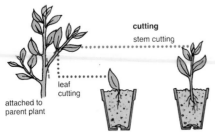

cutting
stem cutting
leaf cutting
attached to parent plant

cutting (*n*) a piece of stem, root, or leaf, cut away from a plant and placed in new soil to produce a new plant by vegetative propagation (↑).

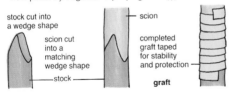

stock cut into a wedge shape
scion
scion cut into a matching wedge shape
completed graft taped for stability and protection
stock
graft

graft (*v*) in plants, to insert a shoot or bud into the stem of another plant so it will continue to grow as a part of the plant. In animals, to transfer, by surgery, a piece of tissue or bone from one part of the body and replace it in another part, where it will grow. **graft** (*n*).

layering (*n*) a branch of a tree, or shrub, is bent and part of it covered with earth in the ground. This branch is kept in place by pegs on either side of a node. Adventitious roots grow from the node. The branch is cut off and transplanted when the roots have formed. Layering is a method of vegetative propagation (↑).

transplant (*v*) in plants, to dig up a plant or seedling and replant it in a new site. In animals, to remove an organ or tissue from one animal and place it in another.

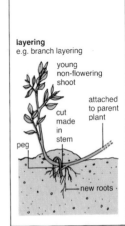

layering
e.g. branch layering
young non-flowering shoot
attached to parent plant
cut made in stem
peg
new roots

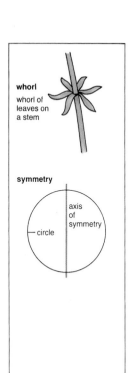

whorl
whorl of
leaves on
a stem

symmetry

axis
of
symmetry

circle

flower (*n*) a structure in monocotyledon (p.230) and
dicotyledon (p.230) plants which includes both
reproductive (called essential) and
non-reproductive (called accessory) parts. It may
have both male and female organs, or the male and
female organs may be borne on separate plants.

corolla (*n*) the coloured petals (↓) of a flower.

whorl (*n*) a circle of similar or identical parts
attached around the stem of a plant at the same
level, e.g. leaves, petals, etc.

perianth (*n*) the outer part of a flower within which lie
the stamens (p.229) and carpels (p.228). It usually
consists of two whorls (↑), an outer one of sepals (↓)
and an inner one of petals (↓).

symmetry (*n*) the state of possessing a regular
shape such that one or more lines exist which divide
the shape into two equal and corresponding halves,
e.g. a line drawn through the centre of a circle
divides it into two equal halves. Such a line is an
axis of symmetry.

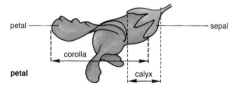

petal — corolla — sepal

petal — calyx

petal (*n*) the inner part of the perianth (↑). In flowers
pollinated (p.229) by insects they are usually
brightly coloured, and exist in different shapes
designed to attract particular insects. Dicotyledons
(p.230) generally have flower parts in groups of five,
monocotyledons (p.230) in groups of three.

sepal (*n*) the outer part of the perianth (↑) in
dicotyledons (p.230), consisting of a green, leaf-like
structure. In dicotyledons they are usually found in
groups of five.

tepal (*n*) in monocotyledons (p.230), a part of a
perianth (↑) which is not clearly differentiated into
sepal and petal, e.g. tulip.

calyx (*n*) a collective name for all the sepals of a flower.

receptacle (*n*) the upper end of the flower stalk in
flowering plants, bearing the sepals, petals, stamens
(p.229) and carpels. In different flowers it can vary in
shape from convex to concave.

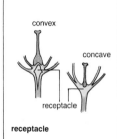

convex

concave

receptacle

receptacle

gynoecium (*n*) a collective name for the female
reproductive (p.225) organs or carpels (↓) of a flower.
It may consist of one or more carpels.

pistil (*n*) alternative name for a **gyanoecium** (↑) or
carpel (↓). **pistillate** (*adj*).

carpel (*n*) the female reproductive organ of a flower,
consisting of a stigma (↓), a style (↓), and an ovary (↓)
containing ovules (↓). There may be one or several
carpels in a flower.

stigma (*n*) the surface of a carpel (↑) on which the
pollen grains are deposited during pollination (↓). The
stigma is usually sticky, so that the pollen grains stick
to it.

style (*n*) the slender, stalk-like part of the carpel (↑)
which supports the stigma (↑). After a pollen (↓) grain
has germinated (p.232), male gametes travel along a
pollen tube, growing down the style, to an ovule (↓) in
the ovary.

nectary (*n*) glands, often at the base of the ovary (↓),
which produce a sugary solution called **nectar**.
Insects are attracted to the flower to drink or collect
the nectar.

ovary (*n*) in flowering plants, the hollow base of the
carpel (↑), surrounded by a thick wall. It contains the
ovules (↓). When fertilized the whole ovary will
develop into a fruit.

ovule (*n*) the part of the ovary (↑) of a flower (p.227)
that contains the female gamete (p.205), and will,
after fertilization, become a seed.

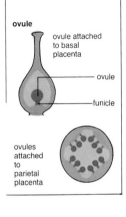

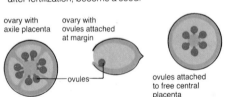

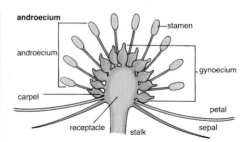

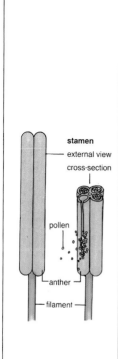

androecium (*n*) a collective name for the male reproductive (p.225) parts of a flower, i.e. all the stamens (↓).

stamen (*n*) the male reproductive part in a flower (p.227); it produces pollen (↓). A stamen consists of an anther (↓) carried on a filament (↓).

filament[2] (*n*) the fine, stalk-like part of a stamen (↑) which supports an anther (↓) in a position suitable to effect cross-pollination (↓).

anther (*n*) the tip of the stamen (↑); it consists of two parts, each containing pollen (↓) sacs.

pollen (*n*) a tiny grain produced by the anther (↑) of a flowering plant. Each pollen grain gives rise to two male gametes. Wind-borne pollen grains are very light. Insect-borne pollen grains are heavier and sticky so they may easily adhere to insects.

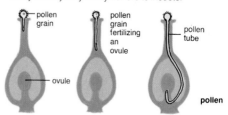

pollination (*n*) the transfer of pollen (↑) from anther (↑) to stigma (↑). The pollen is transported by wind or insects.

cross-pollination (*n*) a type of pollination (↑); the pollen is carried from the anther of one plant to the stigma of another plant.

self-pollination (*n*) a type of pollination (↑) in which pollen is transferred from an anther to a stigma on the same plant.

seed (*n*) in flowering plants, a small structure formed as a result of the fertilization of an ovule (p.228). It contains an embryo ready for germination (p.232) surrounded by a protective coat. Under suitable conditions it will develop and grow into a new plant.

cotyledon (*n*) a simple leaf which forms part of the embryo of a seed. Flowering plants have one or two cotyledons. In some seeds, e.g. pea, bean, they provide a food store for the growing embryo. In others, e.g. grasses, maize, the cotyledon absorbs food from an endosperm (↓) and passes it to the embryo. Within the seed cotyledons do not contain chlorophyll (p.223), but in many plants they grow above ground, develop chlorophyll and carry out photosynthesis (p.223). **cotyledonous** (*adj*).

monocotyledon (*n*) a flowering plant that has only one cotyledon (↑) in its seed. The adult plant bears parallel-veined leaves, has irregularly arranged vascular bundles (p.221), and flower (p.227) parts in multiples of three, e.g. grasses, palm trees. **monocotyledonous** (*adj*).

dicotyledon (*n*) a flowering plant that has two cotyledons (↑) in its seed. The adult plant usually bears leaves with net venation, has flower parts in multiples of four or five, and vascular bundles (p.221) arranged in a ring around the centre of the stem.

endosperm (*n*) food material which surrounds and nourishes the embryo in a seed. Not all seeds have an endosperm. **endospermous** (*adj*).

testa (*n. pl. testae*) the outer, hard protective covering of a seed (↑). It does not let water or oxygen pass when dry, but when wet will allow oxygen to pass through.

hilum (*n*) a scar on the surface of the testa (↑) of a seed at the point where it was attached to the parent plant.

micropyle (*n*) a tiny hole in the testa of a seed through which water enters to help the seed germinate.

embryo (*n*) the part of a seed which grows into a new plant. It is formed from an ovule (p.228) in a flower. **embryonic** (*adj*).

plumule (*n*) part of an embryo (↑). The stem of the seedling develops from it.

shoot (*n*) the stem of a young plant that develops from the plumule; the leaves of the plant grow from it.

radicle (*n*) the part of an embryo which forms the root of a new plant; the first part of a seedling to appear.

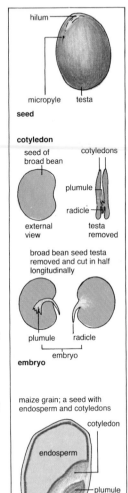

seed

hilum

micropyle   testa

seed

cotyledon

seed of broad bean   cotyledons

plumule

radicle

external view   testa removed

broad bean seed testa removed and cut in half longitudinally

plumule   radicle

embryo

embryo

maize grain; a seed with endosperm and cotyledons

cotyledon

endosperm

plumule

radicle

endosperm

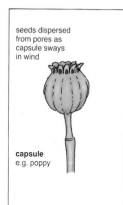

seeds dispersed
from pores as
capsule sways
in wind

capsule
e.g. poppy

dispersal (*n*) the process by which seeds are scattered over a wide area, often very far from the parent plant. Dispersal occurs as a result of the action of wind, water and animals. **disperse** (*v*).

fruit (*n*) the ripened ovary (p.228) of a plant; it contains and protects the seeds. There are many different types of fruit; capsules (↓), drupes (↓), berries (↓), legumes (↓).

capsule (*n*) a dry fruit (↑) containing seeds. When ripe it bursts open to release the seeds, e.g. poppy.

drupe (*n*) a succulent (↓), fleshy fruit (↑) containing an inner stone that holds the seed, e.g. peach, cherry, apricot.

fleshy mesocarp

seed

stalk

epicarp

endocarp or stone

remains of style

drupe
e.g. apricot

berry (*n*) a succulent (↓), fleshy fruit (↑) containing one or many seeds within an inner membrane surrounded by a thick, fleshy wall and a thin, outer skin, e.g. tomato, cranberry.

legume (*n*) (1) a long, narrow, dry fruit, also called a pod. It consists of two halves with the seed fixed along the join; when the pod opens the seeds fall, or burst, out. (2) any plant belonging to the pea-flower family.

succulent (*adj*) describes a plant, stem, leaf or fruit which is soft and thick, and contains a lot of water or sap, e.g. cacti stems are succulent.

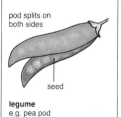

pod splits on both sides

seed

legume
e.g. pea pod

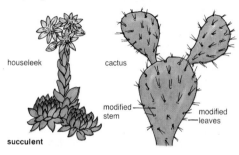

houseleek

cactus

modified stem

modified leaves

succulent

germination (*n*) the start of growth of a seed or spore embryo (p.230), which will develop into a new and independent plant. Germination often follows a period of inactivity, and begins because conditions, such as temperature, or the presence of light, make growth suitable. **germinate** (*v*).

dormancy (*n*) a resting state when plants are not growing. Seeds remain dormant in the soil until the temperature is high enough for them to germinate. Dormancy enables organisms to survive adverse conditions. **dormant** (*adj*).

seedling (*n*) a young plant grown from a seed. The cotyledons (p.230) provide food for the growth of the organism until it becomes capable of manufacturing its own food.

epigeal (*adj*) describes seedlings (↑) whose cotyledons (p.230) grow above ground during germination (↑). Most dicotyledons (p.230) have epigeal seedlings.

epicotyl (*n*) the part of a seedling (↑) or embryo above the cotyledons (p.230).

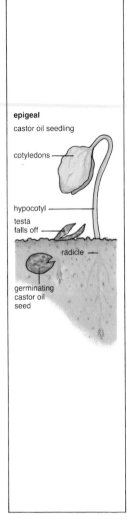

epigeal

castor oil seedling

cotyledons

hypocotyl

testa
falls off

radicle

germinating
castor oil
seed

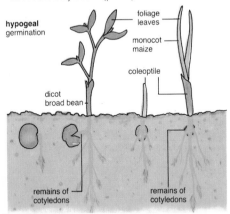

hypogeal
germination

foliage
leaves

monocot
maize

coleoptile

dicot
broad bean

remains of
cotyledons

remains of
cotyledons

hypogeal (*adj*) describes seedlings (↑) whose cotyledons (p.230) grow below ground during germination (↑). Most monocotyledons (p.230) have hypogeal seedlings.

hypocotyl (*n*) the part of the stem below the cotyledons (p.230) and above the root of an embryo or seedling.

some plants will only grow in an acid soil e.g. *Erica*

factor

certain plants thrive in an alkaline, calcareous soil e.g. *Aster*

ecology (*n*) the science concerned with the relations between living organisms and their environment.

flora (*n*) (1) the plant population of a particular area, a specific environment, or a particular period of time. (2) a list of plants, arranged in species, genera, and higher categories, together with a key for identifying the plants. The flora is usually for a particular area.

fauna (*n*) the animal population of a particular area, e.g. the marsupials of Australasia; a specific environment, e.g. the reindeer of the northern tundra regions; a particular period of time, e.g. the dinosaurs of the Mesozoic era.

factor (*n*) one of a number of possible causes that can produce an effect on the plant and animal population of a particular area, a specific environment, or a particular period of time, e.g. the pH value of the soil; the presence of other organisms; the atmospheric humidity, are some of the factors of an environment. *See* **conditions** (↓).

climatic factors the factors (↑) in an environment (↓) which are due to climate, e.g. temperature, rainfall.

biotic factors the factors (↑) in an environment (↓) which are due to the activities of living organisms.

edaphic factors the factors (↑) in an environment (↓) which are due to the physical, chemical and biological characteristics of the soil.

conditions (*n.pl.*) the actual physical and biological items which constitute an environment, e.g. an acid soil (pH Ⓡ 7); the population of predators or consumers in the environment; a tropical climate. Conditions in an environment can be beneficial, tolerable or adverse; they can be necessary or inhibiting. To contrast **conditions** and **factors**: salinity (p.235) is a **factor** of an aquatic environment; a high salinity, as in sea-water, is a **condition** of a particular environment (an ocean).

environment (*n*) all the conditions (↑) surrounding and influencing living organisms (p.143), e.g. atmosphere, light, temperature, other plant and animal life. The environment influences the growth, behaviour and development of organisms.

adaption (*n*) the process by which organisms change and adapt so that they are better suited to their environment, and so increase their chances of survival (p.214). **adapt** (*v*).

mimicry (*n*) the adaption (p.233) of one species to imitate the colour, shape, sound or behaviour of another species or object, e.g. many different species of harmless insects have developed the same marking as those that are poisonous to other animals, and have thereby gained protection from predators.

hibernation (*n*) spending the winter in a dormant (inactive) state. During hibernation the metabolism (p.160) of the animal slows down, the temperature of mammals drops to that of their surroundings, and reptiles and amphibians bury themselves underground.

habitat (*n*) the place where a plant or animal lives, e.g. the sea-shore; a riverbank; scrubland. It may be a region where the organism lives naturally, or a region to which the species has been introduced and has adapted.

aerial² (*adj*) concerned with the air; living or growing in the air, e.g. the parts of a plant that grow above ground.

arboreal (*adj*) concerned with trees; of animals, living in, or adapted to living in trees, e.g. certain apes have arboreal habitats (↑).

terrestrial (*adj*) concerned with the Earth; living or growing on land rather than in water or in the air, e.g. man is a terrestrial animal.

mimicry

mimic

model
poisonous to birds

terricolous

earthworm

terricolous (*adj*) concerned with the soil (p.238); of animals, living in the soil, e.g. a termite is a terricolous organism.

underground, subterranean (*adj*) under the surface of the Earth. Underground is conventionally used to describe objects or organisms immediately below the Earth's surface, e.g. underground stems. Subterranean describes objects or organisms which are very deep in the Earth, e.g. sedimentary rock is a subterranean rock strata.

ground (*n*) the surface of the Earth, in particular the solid surface; the soil or land of the Earth; the foundation of a structure. **ground** (*adj*).

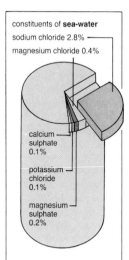

constituents of **sea-water**

sodium chloride 2.8%

magnesium chloride 0.4%

calcium sulphate 0.1%

potassium chloride 0.1%

magnesium sulphate 0.2%

aquatic (*adj*) concerned with water; living and growing on or in the water, e.g. dolphins are aquatic animals; plankton is an aquatic plant.

sea-water (*n*) the water of the seas. It contains different salts, usually in the following concentrations: sodium chloride 2.8%; magnesium chloride 0.4%; calcium sulphate 0.1%; potassium chloride 0.1%; magnesium sulphate 0.2%.

marine (*adj*) concerned with the sea; living in or formed by sea-water, e.g. marine organisms.

salinity (*n*) the amount of sodium chloride (common salt) present in water. Sea-water (↑) usually contains 2.8 g of sodium chloride per 100 g water. Sea-water with a concentration of 4 g sodium chloride per 100 g water has a high salinity. **saline** (*adj*).

brackish (*adj*) describes water which is salty, but less saline than ordinary sea-water.

benthic (*adj*) describes plants and animals living on, or attached to, the sea-bottom. Benthon includes all benthic plants and animals.

pelagic (*adj*) describes organisms which lie on the surface or swim in the open waters of a sea or lake, as opposed to living on the sea- or lake-bed, e.g. plankton and waterlilies are pelagic organisms.

littoral (*adj*) describes organisms living along the shore of an ocean or lake, or on the sea-bed (if shallow, and receiving light). The **littoral zone** of the sea is the area exposed at low tide.

estuary (*n*) the area where the tides enter the mouth of a river. The salinity (↑) of an estuary is highest at high tide and lowest at low tide. The salinity is also highest at the point of the estuary nearest the sea, and lowest at the furthest point up river to which the tide flows.

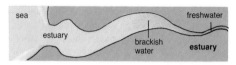

sea　　　　　　　　　　　　　freshwater

estuary

brackish water

estuary

estuarine (*adj*) concerned with an estuary (↑); living in, or formed by, an estuary.

freshwater (*adj*) describes organisms living in water containing no, or very little, salt; concerned with water which contains little, or no, salt, e.g. freshwater ponds, streams or rivers.

biosphere (*n*) the zone of the Earth and its atmosphere (i.e. land, sea and sky) inhabited by living organisms.

low-growing, water-conserving plants

spruce, pines

lichens

coniferous and deciduous trees

biome
e.g. mountain flora

biome (*n*) a regional community of plants and animals covering a large natural area, e.g. savanna; desert; rain forest; mountain.

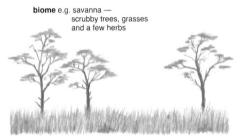

biome e.g. savanna —
scrubby trees, grasses and a few herbs

ecosystem (*n*) a unit formed by a community (↓) of plants, animals and bacteria interacting with each other and with the non-living environment, e.g. forests, riverbeds and cities are different types of ecosystem. Ecosystems contain producers (p.242), mainly green plants; consumers (p.242), omnivores (p.241) and carnivores (p.241); decomposers (p.242), microorganisms which destroy dead organisms.

biotope (*n*) a small region, with the same weather conditions, in which a group of organisms with similar characteristics live together and form a community (↓).

biocoenosis (*n*) a community (↓) of organisms living in a biotope (↑).

territory (*n*) the area of land in which an animal, or a group of animals, lives, feeds or breeds. The limits of the territory are often marked by scent. A territory may be established temporarily for a specific purpose, e.g. mating.

community (*n*) a group of organisms living together in a given area and closely affecting each other, especially through food relationships.

amphibious (*adj*) capable of living on land and in water, e.g. frogs are amphibious organisms.

amphibiotic (*adj*) describes organisms which live in water in one stage of their life cycle (p.145) and on land in another.

motile (*adj*) describes microorganisms which are capable of locomotion, e.g. an amoeba is a motile organism.

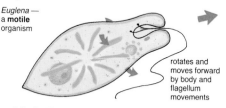

Euglena — a **motile** organism

rotates and moves forward by body and flagellum movements

mobile (*adj*) describes animals which are capable of locomotion (p.189); they are able to move from one place to another.

migratory (*adj*) describes a population of animals which moves from one region to another, usually during a particular season; migratory movements are particularly common amongst birds. Migratory birds fly to a warmer climate during cold seasons of a region when food is less plentiful, and return to the region when temperatures rise. Migratory movements are alternative means of survival to hibernation (p.234) during winter.

sedentary sponge

sedentary (*adj*) describes animals which live fixed to one spot, i.e. not free to move from one place to another, e.g. sponges.

sessile (*adj*) (1) describes plants attached to a support without a stalk (p.221) or stem (p.221). (2) describes sedentary (↑) animals.

diurnal (*adj*) (1) an event that happens daily, e.g. the diurnal rhythm of plants and animals is connected to the cycle of day and night. (2) describes an animal which is active only during daylight.

nocturnal (*adj*) describes an animal which is active only at night, e.g. owls are nocturnal animals.

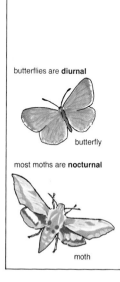

butterflies are **diurnal**

butterfly

most moths are **nocturnal**

moth

soil (*n*) the surface layer of the Earth's crust which provides nutrients (p.166) for plants and animals. It is formed from weathered rocks and contains mineral salts which are dissolved in a fine film of water round each soil particle. The grain structure of soil determines its properties, which in turn determine the types of plants which can grow in it. There are many types of soil, e.g. heavy clays (↓); sandy loams (↓).

topsoil (*n*) the top layer of a soil (↑), above the subsoil (↓), containing the nutrients (p.166) needed by plants for healthy growth. Erosion of the topsoil inhibits or prevents plant growth.

subsoil (*n*) the layers of soil (↑) beneath the fertile topsoil (↑). Subsoil lacks minerals and humus (↓), and is not suitable for growing plants. It can be useful if it prevents water draining away.

soil profile the distinct layers seen in a vertical section through soil (↑). A soil profile usually consists of topsoil and subsoil down to rock.

sand (*n*) soil (↑) comprised of tiny, loose grains of rock, mainly silicon dioxide; water drains through it quickly. **sandy** (*adj*).

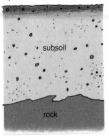

soil profile

topsoil

subsoil

rock

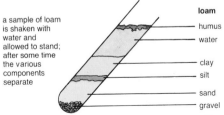

a sample of loam is shaken with water and allowed to stand; after some time the various components separate

loam

humus

water

clay

silt

sand

gravel

loam (*n*) a mixed soil (↑) consisting of fine clays (↓), silt, sand (↑) and humus (↓). It is very fertile, and is the best soil for growing plants.

clay (*n*) a fine-grained soil (↑) composed mainly of particles of aluminium sulphate. The particles are much smaller than sand (↑) particles, and water drains through clay much more slowly than it does through sand.

humus (*n*) a dark brown, organic material found in topsoil (↑), and produced by the decay of plant and animal tissues. It is important for plant growth. Humus contains many nutrients (p.166) and also helps the soil to retain water.

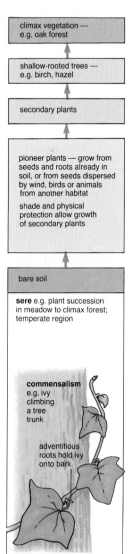

climax vegetation —
e.g. oak forest

↑

shallow-rooted trees —
e.g. birch, hazel

↑

secondary plants

↑

pioneer plants — grow from
seeds and roots already in
soil, or from seeds dispersed
by wind, birds or animals
from another habitat

shade and physical
protection allow growth
of secondary plants

↑

bare soil

↑

sere e.g. plant succession
in meadow to climax forest;
temperate region

commensalism
e.g. ivy
climbing
a tree
trunk

adventitious
roots hold ivy
onto bark

agriculture (*n*) the science of farming; cultivating the soil (↑), growing crops and raising animals to obtain food and other useful materials. Soil conservation (p.243) improves agricultural processes by replacing the nutrients taken out of the soil by plants. Soil erosion must also be prevented.

crop (*n*) any agricultural product grown, harvested or gathered for food or other products, e.g. corn, rice. To conserve the fertility of the land, farmers grow different crops on the same piece of land each year. This is called **crop rotation**.

pest (*n*) any organism that damages crops or spreads disease, e.g. insects, rodents.

weed (*n*) any plant growing where it is not wanted, e.g. growing in a cultivated crop.

escape (*n*) a cultivated crop plant found growing wild due to its seeds being dispersed by wind or animals; it is unwanted and becomes a weed.

manure (*n*) animal faeces and urine mixed with straw bedding or wood shavings, put into the soil to increase its fertility. Manure helps to form humus (↑), and increases the mineral content of the soil.

compost (*n*) a mixture of decayed plant matter put on the soil to increase its humus and mineral content.

fertilizer (*n*) any material, e.g. manure, compost, added to the soil to provide nutrients and improve plant growth. Artificial fertilizers consist of chemical substances which plants use as a source of minerals.

succession (*n*) of plants, the slow series of changes in the species of plants growing in an environment from the types first present to the establishment of a climax (↓). It begins with simple plant life, such as algae and lichens, gradually followed by higher plants, e.g. coniferous and flowering plants.

sere (*n*) a particular type of plant succession (↑) associated with a particular environment.

climax (*n*) the plant community established at the end of a succession (↑) in a particular area, i.e. the final, established vegetation of a region. Provided there are no environmental changes, the climax community will remain unchanged.

commensalism (*n*) a state in which two or more different animals share the same living place or the same food, but are not greatly benefited or harmed by the relationship, e.g. crustaceans growing on the same rock. Contrast **symbiosis** (p.240).

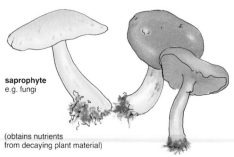

saprophyte
e.g. fungi

(obtains nutrients
from decaying plant material)

saprophyte (*n*) an organism which obtains its food
from dead or decaying plant or animal matter,
e.g. fungi, bacteria. Saprophytes play an important
role in completing the carbon and nitrogen cycles
(p.244). **saprophytic** (*adj*).

epiphyte (*n*) a plant growing on another plant, and
using it for support. An epiphyte does not get food
from the support, i.e. it is not a parasite (↓).

symbiosis (*n*) a state in which two different
organisms live together in a close relationship
which is mutually advantageous, e.g. the
nitrogen-fixing bacteria in the root nodules of
leguminous (p.231) plants; the bacteria provide
nitrates for the plant, the plant provides nutrients
(p.166) for the bacteria.

symbiont e.g. lichen
(algae and fungi
symbiotically living together)

symbiont (*n*) one of the two organisms living
together in symbiosis (↑).

mutualism (*n*) an alternative name for **symbiosis** (↑).

parasite (*n*) an organism which lives inside or on the
surface of another living organism, the host (↓), from
which it obtains food, and sometimes shelter,
without benefiting the host and usually harming it.

host (*n*) an organism which provides a parasite (↑)
with food and protection.

epiphyte
e.g. orchid
(found growing in tropical
forests on horizontal
branches; the orchid's roots
take up water and
nutrients from soil and plant
debris built up in the crevices
of bark)

nekton (*n*) all the free-swimming animals of the seas or freshwater, e.g. whales, dolphins and salmon are part of nekton.

plankton (*n*) microscopic plants and animals which float near the surface of the sea and freshwater lakes. The plants use the light at the surface for photosynthesis (p.223). Plankton is an important source of food for many aquatic (p.235) organisms, and is usually at the beginning of a food web (p.166) of aquatic organisms.

phytoplankton (*n*) the microscopic plants of plankton (↑). They are the primary source of food in an aquatic (p.235) food web (p.166), providing food for zooplankton (↓) and larger organisms, such as fish and whales. Phytoplankton are not capable of locomotion, but float or drift.

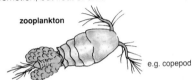

zooplankton

e.g. copepod

zooplankton (*n*) the microscopic animals of plankton. They feed on phytoplankton (↑), and most use flagellae (p.153) for locomotion. They act as a food source for fish.

herbivore (*n*) an animal that feeds only on plants, e.g. sheep are herbivores. **herbivorous** (*adj*).

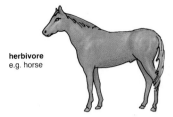

herbivore
e.g. horse

carnivore

predator e.g. eagle

prey e.g. mouse

carnivore (*n*) an animal that feeds only on other animals, e.g. an eagle. **carnivorous** (*adj*).

predator (*n*) an animal that hunts and eats other animals. **predatory** (*adj*).

prey (*n*) an animal hunted and eaten by a predator.

omnivore (*n*) an animal that eats plants and animals, e.g. humans are omnivores. **omnivorous** (*adj*).

producers (*n.*) organisms which manufacture their own food. Producers are normally green plants which use inorganic (p.112) materials in photosynthesis (p.223) to produce proteins (p.163), carbohydrates (p.164) and fats (p.165). All food chains (p.166) and food webs (p.166) begin with a producer.

autotroph (*n*) an organism which builds up organic food materials from inorganic compounds and thus does not depend on outside sources to provide organic substances as nutrients. The energy used to build up these food materials is obtained independently of outside substances, e.g. as in photosynthesis (p.223). Green plants and a few species of bacteria are examples of autotrophs. **autotrophic** (*adj*).

consumers (*n.*) an organism which feeds on other organisms or their products. Consumers cannot manufacture their own food. **Primary consumers** are herbivores (p.241), **secondary consumers** are carnivores (p.241).

decomposers (*n.pl.*) microscopic organisms which cause the decay of dead plant and animal tissues to form inorganic (p.112) materials, completing the carbon and nitrogen cycles. Fungi (p.142), bacteria (p.143) and some protozoa (p.140) are decomposers.

trophic level
trophic levels in a food chain
— carnivores
— herbivores
— plants

trophic level one of the stages into which a food chain (p.166) may be divided. Plants are at the first trophic level; herbivores (p.241) are at the second trophic level; carnivores (p.241) form higher trophic levels.

pyramid of numbers the number of organisms at each trophic level (↑) in a food chain (p.166) or food web (p.166). Energy is lost in going from one trophic level to a higher one because of respiration, heat radiation (p.98) and other metabolic (p.160) processes. So, at higher levels, less energy is available, and usually the organisms are larger; therefore, fewer organisms are found at higher trophic levels than at lower levels.

producer
e.g. grass

size of animal

large carnivore

small carnivore

herbivore

producer

number of individuals

pyramid of numbers

conservation (*n*) the preservation and careful use of natural resources, such as seas, rivers, countryside. Conservation entails replacing used resources, such as soil nutrients or fish stocks; or ensuring that limited resources, such as coal or oil, are used carefully.

resource (*n*) the supply of plants, animals, fuels, minerals and other raw materials that are available for use by a country, or region. Resources also include productivity of the soil and means of generating power from natural forms of energy such as hydroelectric power.

recycle (*v*) to take used materials and treat them chemically and physically so that the original material is reformed and can be used again, e.g. to recycle paper by bleaching it, pulping it, and producing new paper.

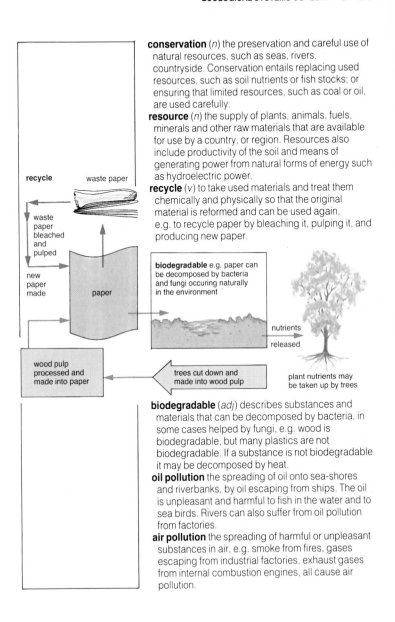

recycle   waste paper

waste paper bleached and pulped

new paper made

paper

biodegradable e.g. paper can be decomposed by bacteria and fungi occuring naturally in the environment

nutrients released

wood pulp processed and made into paper

trees cut down and made into wood pulp

plant nutrients may be taken up by trees

biodegradable (*adj*) describes substances and materials that can be decomposed by bacteria, in some cases helped by fungi, e.g. wood is biodegradable, but many plastics are not biodegradable. If a substance is not biodegradable it may be decomposed by heat.

oil pollution the spreading of oil onto sea-shores and riverbanks, by oil escaping from ships. The oil is unpleasant and harmful to fish in the water and to sea birds. Rivers can also suffer from oil pollution from factories.

air pollution the spreading of harmful or unpleasant substances in air, e.g. smoke from fires, gases escaping from industrial factories, exhaust gases from internal combustion engines, all cause air pollution.

cycle (*n*) a continuous, regularly repeating series of events. **cyclical** (*adj*).

carbon cycle the cycle (↑) of carbon, a constituent of all organic compounds and all organisms. Carbon is essential for life. During photosynthesis (p.223), plants use carbon dioxide from the atmosphere to synthesize (p.116) organic (p.113) compounds. The carbon is circulated in one of three ways: carbon compounds are passed on to animals which eat the plants; carbon dioxide is released by the plant during respiration; carbon compounds are oxidized to carbon dioxide by microorganisms feeding on dead or decaying plants. Animals which obtain carbon compounds by eating plants also release carbon dioxide back into the atmosphere by respiration, or decay after death. The carbon dioxide is used by plants during photosynthesis, and the cycle begins once more.

nitrogen cycle the cycle (↑) of nitrogen, a constituent of all proteins (p.163). Plants make proteins from nitrates (p.112) in the soil. The plants either decay, returning nitrogen to the soil, or are eaten by animals which use the plant proteins. Nitrogen is returned to the soil by the decay of the animals, or in animal excreta as urine (p.183). The action of bacteria (p.143) on the ammonia (p.183) set free from urea (p.183) produces nitrates and protein. This completes the cycle.

water cycle the cycle (↑) of water in nature. Rain falling from clouds provides water for the soil and plants. Excess water drains into streams and rivers, and is used by animals, as it flows to the sea. Plants, animals, and exposed water surfaces give off water vapour to the atmosphere, where it forms clouds. Clouds deposit rain and the cycle continues.

soft water water which does not contain magnesium or calcium salts and easily forms a lather with soap.

hard water water which contains salts of magnesium, calcium or both metals. It does not easily form a lather with soap. Instead, the salts combine with the soap to form **scum** (solid particles on the surface of the water). Temporary hard water can be softened by boiling. Permanent hard water can be softened by adding sodium carbonate.

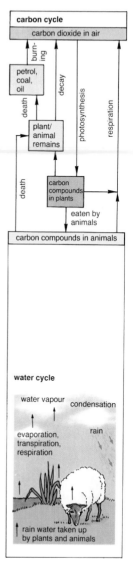

The periodic table classifies the elements according to their atomic number. The elements are arranged in groups (arranged vertically) or periods (arranged horizontally). There is a scientific connection between elements within one group or period, e.g. the elements of fluorine, chlorine and bromine are members of the group of elements called halogens; they are some of the most highly reactive elements.

KEY

HYDROGEN

ALKALI AND ALKALINE EARTH METALS

METALS

NON METALS INCLUDING HALOGENS

NOBLE GASES

H -																	He 4
LITHIUM Li 7	BERYLLIUM Be 9											BORON B 11	CARBON C 12	NITROGEN N 14	OXYGEN O 16	FLUORINE F 19	NEON Ne 20
SODIUM Na 23	MAGNESIUM Mg 24											ALUMINIUM Al 27	SILICON Si 28	PHOSPHORUS P 31	SULPHUR S 32	CHLORINE Cl 35.5	ARGON Ar 40
POTASSIUM K 39	CALCIUM Ca 40	SCANDIUM Sc 45	TITANIUM Ti 48	VANADIUM V 51	CHROMIUM Cr 52	MANGANESE Mn 55	IRON Fe 56	COBALT Co 59	NICKEL Ni 59	COPPER Cu 64	ZINC Zn 65	GALLIUM Ga 70	GERMANIUM Ge 73	ARSENIC As 75	SELENIUM Se 79	BROMINE Br 80	KRYPTON Kr 84
RUBIDIUM Rb 85	STRONTIUM Sr 88	YTTRIUM Y 89	ZIRCONIUM Zr 91	NIOBIUM Nb 93	MOLYBDENUM Mo 96	TECHNETIUM Tc 98	RUTHENIUM Ru 101	RHODIUM Rh 103	PALLADIUM Pd 106	SILVER Ag 108	CADMIUM Cd 112	INDIUM In 115	TIN Sn 119	ANTIMONY Sb 122	TELLURIUM Te 128	IODINE I 127	XENON Xe 131
CAESIUM Cs 133	BARIUM Ba 137	LANTHANUM La 139	HAFNIUM Hf 178	TANTALUM Ta 181	TUNGSTEN W 184	RHENIUM Re 186	OSMIUM Os 190	IRIDIUM Ir 192	PLATINUM Pt 195	GOLD Au 197	MERCURY Hg 201	THALLIUM Tl 204	LEAD Pb 207	BISMUTH Bi 209	POLONIUM Po 209	ASTATINE At 210	RADON Rn 222
FRANCIUM Fr 223	RADIUM Ra 226	ACTINIUM Ac 227															

Some useful abbreviations and constants

Common abbreviations

abs.	absolute	insol.	insoluble
a.c.	alternating current	i.r.	infrared
anhyd.	anhydrous	liq.	liquid
a.p.	atmospheric pressure	m.p.	melting point
aq.	aqueous	p.d.	potential difference
b.p.	boiling point	ppt.	precipitate
conc.	concentrated	r.a.m.	relative atomic mass
concn.	concentration	r.d.	relative density
const.	constant	sol.	soluble
crit.	critical	soln.	solution
cryst.	crystalline	s.t.p.	standard temperature and pressure
d.c.	direct current		
dil.	dilute	temp.	temperature
dist.	distilled	u.v.	ultraviolet
e.m.f.	electromotive force	v.d.	vapour density
f.p.	freezing point	V.R.	velocity ratio
h.	hour	w.	weight
hyd.	hydrated		

Physical constants

PHYSICAL CONSTANT	SYMBOL	VALUE
Avogadro constant	No	$6.02 \times 10^{23} \, mol^{-1}$
atomic mass unit	a.m.u.	$1.660 \times 10^{-27} \, kg$
proton mass	1.007 a.m.u.	$1.673 \times 10^{-27} \, kg$
neutron mass	1.009 a.m.u.	$1.675 \times 10^{-27} \, kg$
gas constant		$8.314 \, JK^{-1} \, mol^{-1}$
electron-volt	eV	$1.60 \times 10^{-19} \, J$
electronic charge	e	$1.60 \times 10^{-19} \, C$
Faraday constant	F	$9.65 \times 10^{4} \, Cmol^{-1}$
Planck's constant	h	$6.62 \times 10^{-34} \, Js$
one calorie	C	$4.18 \, J$
electron mass	m	$9.11 \times 10^{-31} \, kg$
s.t.p.		1.00 atm or 760 mm Hg or 101 KPa or 0°C or 273.15 K
temperature triple point water		273.16 K
standard volume of mole of gas at s.t.p.		$22.4 \, dm^3$
specific heat capacity water		$4.18 \, Jg^{-1} K^{-1}$

Index